SUDDENLY AN ENGLISHMAN

SUDDENLY AN ENGLISHMAN

'THE LIFE OF LOUIS HAGEN' AND 'ARNHEM LIFT, A GERMAN JEW IN THE GLIDER PILOT REGIMENT'

LOUIS HAGEN

First published in this combined edition, 2024
Arnhem Lift first published 1945

The History Press
97 St George's Place, Cheltenham,
Gloucestershire, GL50 3QB
www.thehistorypress.co.uk

© Louis Hagen, 1945, 2012, 2024

The right of Louis Hagen to be identified as the Author
of this work has been asserted in accordance with the
Copyright, Designs and Patents Act 1988.

All rights reserved. No part of this book may be reprinted
or reproduced or utilised in any form or by any electronic,
mechanical or other means, now known or hereafter invented,
including photocopying and recording, or in any information
storage or retrieval system, without the permission in writing
from the Publishers.

British Library Cataloguing in Publication Data.
A catalogue record for this book is available from the British Library.

ISBN 978 1 80399 719 3

Typesetting and origination by The History Press
Printed and bound in Great Britain by TJ Books Limited, Padstow, Cornwall

 Trees for LYfe

Contents

The Life of Louis Hagen

Foreword   9

1. London, January 1936   11
2. My Family   17
3. My Early Life   25
4. Schloss Lichtenberg   31
5. Making Friends   37
6. On and Off the Factory Floor   41
7. A Salesman in London   47
8. The Phoney War   55
9. The Pioneer Corps   59
10. Volunteering   63
11. Learning to Fly   71
12. Arnhem   77
13. The Military Medal   93
14. India   101
15. Becoming a Journalist   113
16. Returning to Germany   125
17. A New Beginning   133

Epilogue   149

Arnhem Lift

Acknowledgements		163
Prefatory Note to the First Edition		165
Foreword to the Second Edition		167
1	About the Author	169
2	The Background	179
3	Arnhem Lift	183
	Monday	183
	Tuesday	188
	Wednesday	196
	Thursday	200
	Friday	204
	Saturday	223
	Sunday	241
	Monday	247
4	Winrich Behr's Story	263
5	The German View of the Battle	271
	Sunday	271
	Monday	277
	Tuesday	279
	Wednesday	280
	Thursday	280
	Friday	282
	Saturday	284
	Sunday	285
	Monday	285
	Tuesday	286
6	Summing Up	289
7	Life After Arnhem	293
	Index	299

THE LIFE OF LOUIS HAGEN

LOUIS HAGEN

Foreword

I must admit that I am ashamed it has taken me so long to get this autobiography published. I have no excuse other than that while my father, Louis (known familiarly as Büdi), was still alive, we sent this manuscript to many publishers who turned it down on the grounds that the market was flooded with Holocaust survivor stories. It was only when I realised *Arnhem Lift* was out of print that I contacted The History Press to see if they were going to reprint it, and happened to ask them if they would be interested in publishing my father's autobiography alongside it. I also mentioned that it was the eightieth anniversary of the Battle of Arnhem, and they came up with the idea of publishing this autobiography alongside *Arnhem Lift*. All I have done is add photographs and corrected some text – otherwise it is exactly as Büdi wrote it.

An awful lot has happened to my family since. I am now an orphan (although I do have a wonderful family of my own) as both my mother and sister Siri have since died, which could be the subject of another book in and of itself.

But this is Büdi's story. I think the reader will sense from reading this that he was a very lucky and very brave man in so many ways, but I don't think it was just luck so much that he survived – I think it's because he was always such a '*mensch*'. He held no grudges towards what he called

'The German Disease', despite being imprisoned in a concentration camp at the age of 16.

As soon as the war was over, Büdi felt the need to return to Germany to try to understand why swathes of his country chose to follow Hitler. As a result he wrote *Follow My Leader* (also published as *Mark of the Swastika*), which was then later republished by The History Press as *Ein Volk Ein Reich: Nine Lives Under the Nazis*. It has achieved critical acclaim as one of the best books written explaining how and why Hitler rose to power.

My father's luck continues to this day: after the fall of the Berlin Wall my cousin, Dr Louis Hagen (a lawyer living in Germany), spent many years working on, and still is, the restitution of the Hagen family estate, which has transformed the lives of all the surviving family members scattered around the world in London, Munich and New York.

I am eternally grateful to have had Büdi in my life. While he was not perfect, I could not really have wanted for a kinder, wiser or more generous father.

Caroline Hagen Hall.
England, February 2024

1

London, January 1936

Even today, after so many years, I can still remember exactly how I felt as the train rattled over the rails on the last stage of my journey from Potsdam to London. I was 19 and leaving home because it was no longer possible for Jews to live safely in Germany. I had already learnt what persecution meant – I had spent six weeks in a concentration camp. As the train drew into Victoria Station, I felt a mixture of joy and relief. Here in England, I was to be an ordinary human being, not different from other people, not a second-class citizen or an outsider because I was a Jew.

Since I had left Germany, I had seen no black or brown uniforms, and no stiff-armed Hitler salutes. When I climbed out of the carriage with my bags and looked around, it gave me real pleasure to not see swastika flags, armbands or badges. Instead, waiting to meet me and smiling, was Dr Lothar Mohrenwitz, an old family friend. We children had always called him Mumpitz, meaning nonsense in German, because of his bubbling good humour and fund of silly stories. He was grey haired now and seemed to have shrunk in the two years since we had last seen each other. He hugged me and said, 'My, how you've grown!'

Mumpitz took me back to the flat in Fitzroy Square that he shared with the drama critic Hubert Griffith and his girlfriend, Kay. I was content just to sit and listen to them, and didn't feel I needed to talk. This was just as well, as it soon became clear that my English wasn't as good as I had been led to believe.

Mumpitz saw me up the stairs to a little room containing a bed, a bookcase, a wardrobe and not much else. I soon fell into an exhausted sleep.

But not for long. It seemed only a few moments later that the door opened and someone started undressing by the dim light from the landing. At first I was scared, then I reasoned that a burglar was unlikely to undress.

I cleared my throat to make my presence known. A burly figure peered threateningly at me and growled, 'What are you doing here? Don't you know this is my room when the House is sitting? Everybody knows that!'

His words made no sense to me. I pretended to go back to sleep and, grumbling to himself, the intruder gathered up his clothes and lumbered out of the room.

Early next morning, when I crossed the hall to the lavatory, I saw a large bundle uncomfortably curled up on a small sofa. The bundle turned out to be Nye Bevan, an up-and-coming young Labour MP. At breakfast, he brushed aside my apologies and bombarded me with questions about my experiences in the concentration camp, my background, and about the country I had left behind.

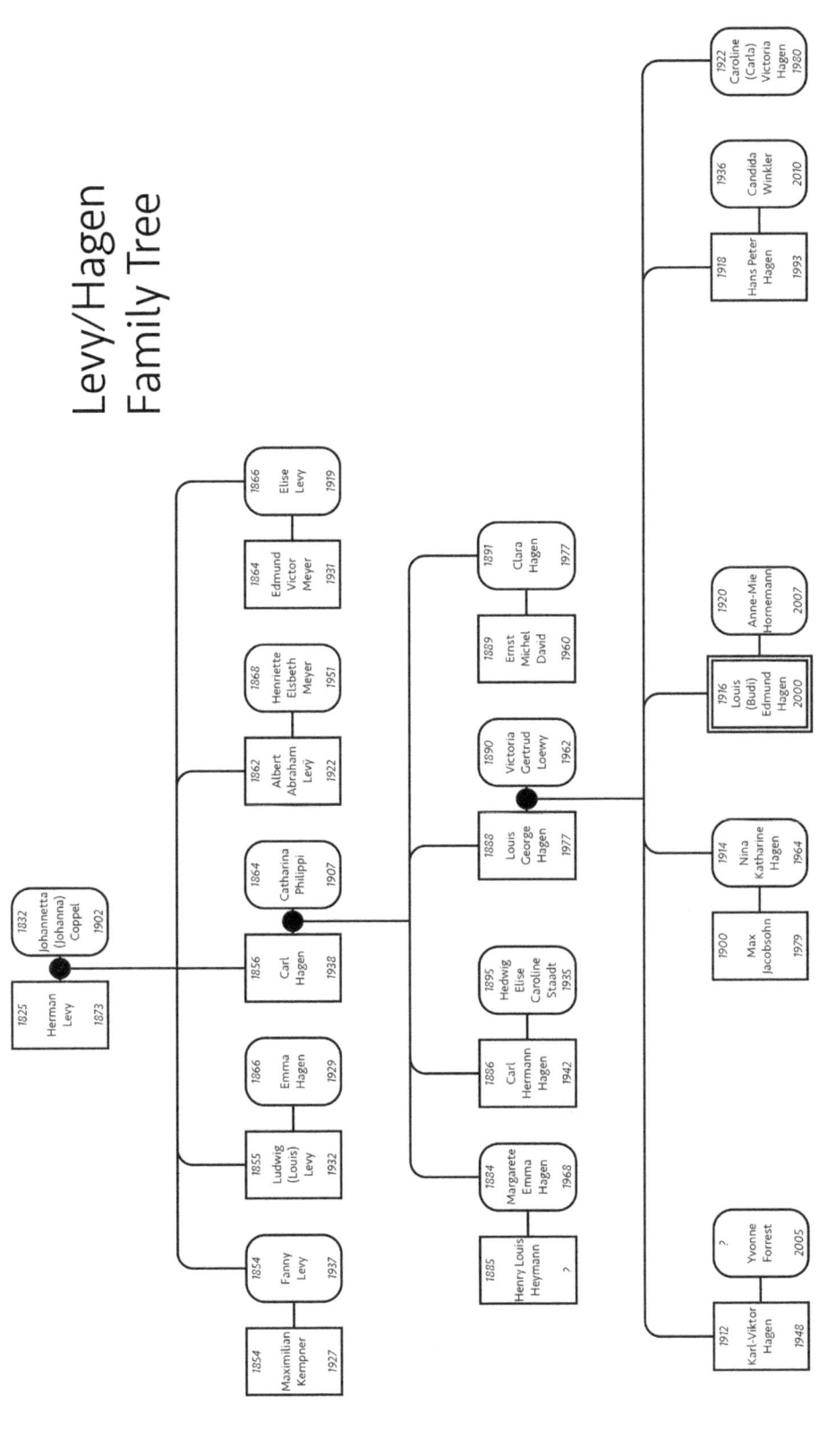

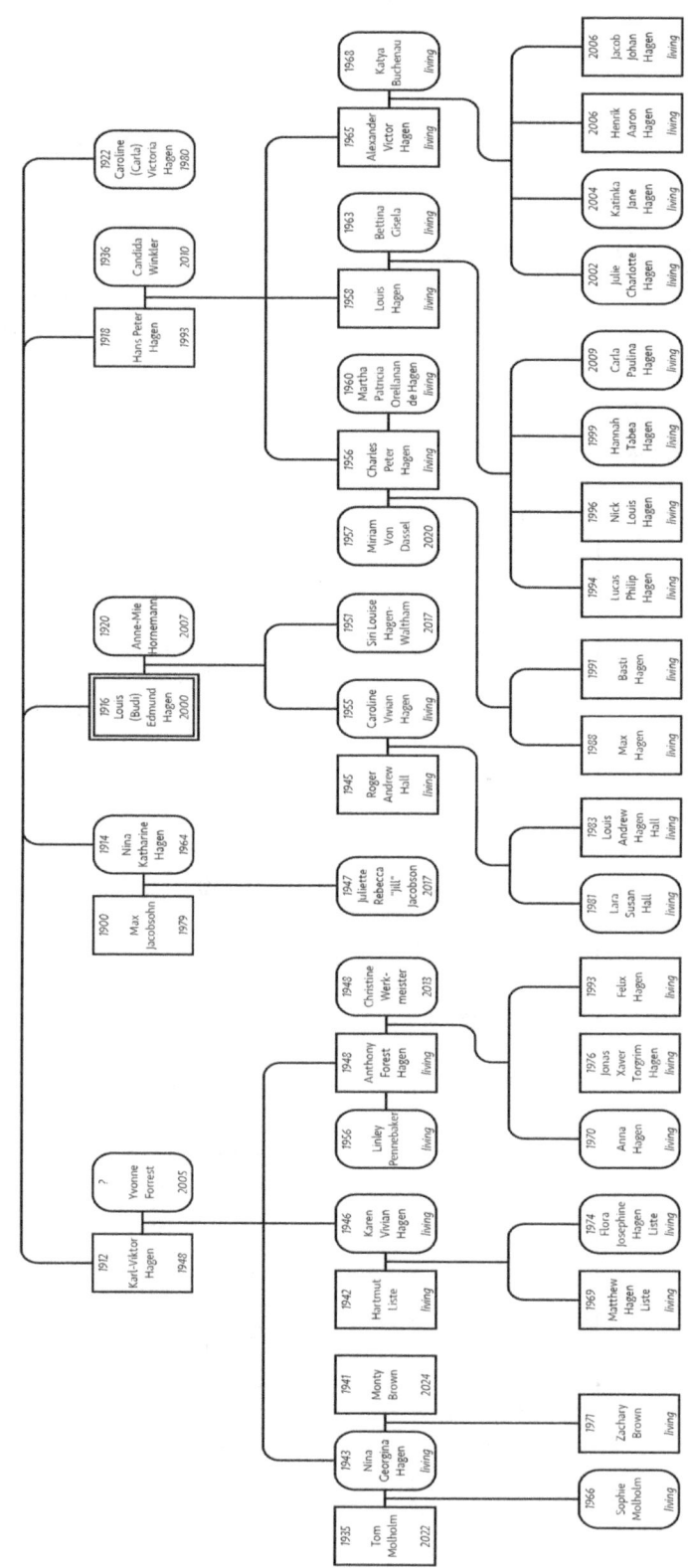

The Loewy children. Louis's mother, Victoria Gertud, is second from right. She emigrated to America via Japan on 11 November 1940. The boy on the far right is Curt Edgar, who died in Gurs concentration camp.

2

My Family

My family, originally called Levy, had been bankers in Germany for five generations. At an early age, Abraham Levy, born at the end of the eighteenth century, became a messenger for the well-known merchant bank of Salomon Oppenheim in Cologne. At the time, it was almost impossible for a lowly employee to rise through the ranks, but young Abraham was discreet and friendly. He was soon noticed by many rich and influential customers who helped set him up in business, first as a registered stockbroker, then as a banker. His son, Herman Levy, born in 1825, expanded the business, and in 1858 turned it into the family bank, A.E. Levy, with the help of the dowry from his wife, the daughter of a wealthy Jewish steel manufacturer.

Their oldest son, Louis, was to build on the bank's success, becoming a leader of Rhineland industry and president of the Cologne Chamber of Commerce for seventeen years. A well-known *bon vivant*, he was immensely rich, had an elegant house in the most fashionable part of Cologne, a great country mansion, and a number of beautiful mistresses, each in her own luxurious apartments. After the end of the First World War, he and an ambitious young politician, Konrad Adenauer, led a movement to create an independent Rhineland state that would be closely linked to France. Louis was to be the finance minister and the currency unit, so it was rumoured, was to be the Louis D'Or.

This ambitious scheme failed to get off the ground. Nevertheless, the bank continued to prosper, even when many others were failing. In 1923, A.E. Levy merged with Salomon Oppenheim, and Louis became its head; so the grandson of the bank messenger became its chairman. Salomon Oppenheim remains one of Europe's leading private banks, and Louis's descendants, now married into the German aristocracy, are still part of it.

In 1896, when Louis Levy married Emma Hagen, the daughter of a wealthy Catholic industrialist, he changed his surname to hers. His younger brother, Karl, my grandfather, had also changed his name to Hagen and was by now doing very well with his own bank in Berlin, Hagen & Co. He was on the board of thirty-four companies and had built a mansion in Berlin and a large country house on the outskirts of the city, in the fashionable small town of Potsdam.

I never met my grandmother, Katherina, but I still have a lovely portrait of her aged 17, painted in 1880 by the well-known Berlin artist Karl Gussow. My grandfather had seen the picture in the Berlin annual art exhibition and was fascinated by the young girl with the soft dark hair and long plait. He fell in love with the picture, traced her parents' address, and found that her father was a colleague of his on the Berlin Stock Exchange. Katherina, just as beautiful in real life, was by no means overwhelmed by the idea of marrying before she'd had a chance to enjoy life. Her parents, however, were impressed by the suitability of my grandfather as a son-in-law, and finally it was agreed that the pair should be married when Katherina reached her 18th birthday. It was by all accounts not a happy marriage, but she bore him two daughters and two sons – one of them my father – before her death from breast cancer when my father was still in his teens.

Soon after, Karl fell in love with Julia Wettstein, a striking young woman who was in charge of the lingerie department in one of Berlin's most exclusive stores – which he later bought for her, and to which I remember being taken as a boy by my mother. One Sunday in 1925, about fifteen years after Katherina's death, there was a big family gathering at my grandfather's Potsdam mansion. To my astonishment, everybody stood up and drank champagne (we children had raspberry juice), toasting the health and happiness of Frau Wettstein and my grandfather, who had just got married. Many years later, I asked my father why Opapa (German for grandfather) had not married Julia earlier. Opapa, he explained, had felt it was not proper to bring into the family unit a young wife from an 'inferior' social background, until all the children, especially the girls, had left home. Social standards were very strict before the First World War, especially for the likes of a private banker.

Julia made my grandfather very happy. She called him Charles, and he called her Oiseau, his little bird. I can still remember them holding hands and looking into each other's eyes like a young couple in love. Though petite and apparently fragile, Julia was the most efficient and energetic housekeeper he could have found for the large establishments he kept in Berlin and Potsdam. The domestic routines ran as smoothly and unobtrusively as those in a five-star hotel. Most of the servants stayed for life, and I can still remember most of their names.

Herr Kotter was just about Opapa's age and had been his coachman before the First World War. By the time I knew him, he had been promoted to chauffeur and was in charge of the enormous supercharged Mercedes, which he endlessly polished with great love and pride. My grandparents took long trips in it to different spas in Germany, and once drove to Nice, at the time quite an undertaking. On these journeys, Kotter, who was getting on, was assisted by the second butler, Arthur, who read the map or took over the wheel when the old man got tired. Our house flag was light and dark blue. My grandfather was an Anglophile and an enthusiastic follower of the Oxford–Cambridge boat race. To avoid taking sides, he adopted the colours of both universities. The chauffeurs wore light blue dust coats with dark blue lapels and cuffs, and matching peaked caps. Opapa also had a specially built town car, high enough for him to get in without taking off his hat.

On Wilhelm Kotter's fiftieth anniversary in Opapa's service, in the late 1920s, my grandfather gave him his own little car. It was an Austin Seven built under licence in Germany by BMW, which he had set up in 1923 in partnership with the Deutsche Bank. Articles about Wilhelm appeared in all the Berlin newspapers – a chauffeur with his own car was unheard of at the time. At least once a week, my bulky grandfather squeezed into the baby car to give Wilhelm the pleasure of driving him to the bank or the stock exchange in his own car.

Each summer, our whole family, with any friends who happened to be staying, often spent the day at Opapa's house in Potsdam. Some of us went by motorboat, crossing the town through an intricate system of canals; some went by car, but most of the children went on bicycles. Opapa received the children sitting at the electric pianola in the entrance hall, where he pretended to play some wild Tchaikovsky piece that invariably ended with his falling off the piano stool. Every time, we shrieked with laughter and pleaded with him to play some more. We could not get enough of the outrageous stories he used to invent about our mythical ancestors: Pinechens von Bonn, Pulm von Birchen and Gedale–Dalje–Leibje. At the time, we believed everything he said.

These characters also appeared in a review we put on for his 70th birthday, written in verse by a well-known playwright and accompanied by popular songs. Opapa was a generous and public-spirited man. Many of his gifts of paintings, by Renoir, Monet, Manet and others, still hang in the Berlin National Gallery, and he often took his grandchildren to admire them. He also thought it good for our education to go to the opera, where he had his own box. Unfortunately for his theories, I found the visits to the gallery and opera a bore, and they put me off the arts for years.

My other grandfather died quite young, long before I was born, at the turn of the century. Photographs show a tall, slim, elegant man with dark, wavy hair and an uncanny likeness to Lord Byron. A successful entrepreneur, he had taken advantage of the explosive development of Berlin after it became the capital of the newly created German Reich and the seat of the Kaiser. Besides other properties, he owned the smart Hotel Eden and a six-floor apartment house near the Kurfürstendamm, one of the two main shopping boulevards of Berlin. There my grandmother had a spacious, twelve-roomed maisonette in which to bring up their three boys and four girls, of whom my mother, Vicky, was the youngest and prettiest.

My father had known Vicky for many years because she went to the same school as his sisters. Then they had worked together in Die Zentrale, a charity founded by his Uncle Albert to look after the widows and orphans of bank employees. My father had been in love with her for as long as he could remember, but she wanted to learn about life and have a good time before getting married and bearing the inevitable succession of children. So, my father left for America to gain business experience in a New York bank.

Villa Carlshagen, Louis' grandfather's house, Potsdam.

Dining room, Carlshagen.

All the grandchildren, Carlshagen, *c*.1927. Louis is standing on the left, back row.

Carl Hagen, *c.*1930.

The wedding day of Louis' parents, Louis Georg and Vicky Hagen, 1912.

Louis' parents at Bertinistrasse, Potsdam, 1935.

3

My Early Life

On my father's return from America in 1911, he took advantage of Vicky's obvious pleasure in seeing him again to seduce her – an almost unbelievable feat at the time. He was also years ahead of his time in arranging for her to be fitted with a vaginal cap, so they could avoid the danger of pregnancy and discovery of their relationship outside of wedlock. If their guilty secret had got out, both families would have become social outcasts. They were married in 1912 and, like her sisters, Vicky was given a dowry of half a million gold marks, a substantial sum at the time.

During the First World War, my father served as a naval officer in the Baltic. His was probably one of the fastest promotions in the German Navy, from civilian to captain of a ship in a single week. My grandfather had used his connections with the German Admiralty in Berlin to propose that, as the navy was short of patrol vessels, he would loan his 30m yacht, *Matz*, for the duration of the war, on the condition that my father commanded it. The Admiralty agreed, and that was how my father got safely and unheroically through the First World War.

His most notable exploit during this time involved ramming his own admiral's barge. The naval base was at Kiel, but whenever the *Matz* returned from patrol, my father disembarked earlier, at the little seaside resort of Travemünde where his family were staying. One evening, my father's ship scraped the admiral's barge on its way into dock.

My father was not on board because he was already safely in bed with my mother in the Grand Hotel Travemünde. This was a crime for which he should have been court-martialled but, again through my grandfather's influence, he was given an alternative: if he volunteered for service on the Western Front, the whole matter would be forgotten. He took the alternative.

At the front, the military commander had no idea what to do with a totally inexperienced naval officer whose only asset was his heroism in volunteering for service on the front line. However, my father's knowledge of horses and boats landed him in charge of a string of horse-drawn barges on the Belgian canals for a few months, until his father managed to get him returned to the *Matz*.

After the war, my grandfather, tired of the day-to-day running of the bank, asked my father to take over its management. He accepted, with reluctance, and remained an unenthusiastic banker all his life. He was always much more interested in technology, the arts, philosophy and history, than in power or making money. But make money he did; it was very difficult to own a private bank at the time and not do so.

Though not personally interested in luxury – he was spartan in his personal life, hated waste and always tried to save money on even small expenses – my father regularly overspent his income on expensive hobbies. He built a huge super-modern house and a private school for his children and their friends; he made films that never earned a fraction of their production costs, and financed all sorts of avant-garde artistic ventures. He was most generous when large sums and projects were involved, and because of this, he was nearly always in debt and, it seemed to us children, worried about money.

We were five children: Katheriena Herta (Nina), born in 1912; Karl-Viktor (always called KV), born in 1914; me, Louis Edmund, born in 1916; Hans Peter, born in 1918; and Karoline (Carla), the youngest, born in 1922. I was always called Büdi, a mispronunciation by KV and Nina of *Bruderlein*, meaning little brother. We should have been six, but my mother had a miscarriage in 1920.

In 1925, my father commissioned the architect Otto Block to build one of the first modern Bauhaus-style houses on the Jungfernsee, a lake just outside Potsdam where Prussian royalty had established their summer palaces. The neighbouring house was the summer residence of the Crown Prince; it was later used for the Potsdam Conference between Churchill, Stalin and Truman.

It was a marvellous place to grow up. The large main hall was decorated with Dutch landscapes and an elegant fireplace. At the far end, mirrored sliding doors led on to a wide terrace, with lovely views across the lake to where distant barges and steamers made their way towards the River Elbe, the port of Hamburg and the North Sea. We had twelve roof terraces, our own cinema and a gymnasium that included a boxing ring and shower room. The walls of my mother's sitting room were covered in ivory silk, hung with a collection of miniatures. The music room had a Bechstein grand piano and space for a large audience. From here, a low step led through heavy damask curtains into a low-ceilinged room with Persian carpets, sofas all around the walls and eastern-style furnishings. We called it The Harem and it was the perfect place for our juvenile attempts at seduction. The English Regency dining table could seat up to eighteen people, as could the huge square oak table in our children's dining room, which was just as well, as the house was often filled with guests and relations – Mumpitz among them – especially during the summer holidays.

These guests happily joined my parents in a lifestyle that was, for those days, very progressive. Our diet was almost wholly vegetarian, with all the vegetables home grown, and on one day each week we ate only raw food. We children longed for meat and sometimes managed to persuade one of the guests to take us out secretly for a proper meal.

Early every morning, Fräulein Ulrich would turn up with her tambourine to conduct naked gymnastics on the lawn, which stretched down to the lake in front of the house. This nudism once got my father into trouble. He and Nina were having an early morning swim, of course without bathing suits, when, to his astonishment, abusive shouting came from the owner of a yacht anchored close to our landing place. 'Disgusting! Perverted! Irresponsible!'

My father mildly pointed out that, if the yacht owner objected to nudism on private property, he was at liberty to sail away and moor elsewhere. The abuse continued, loud enough to wake the sleeping household. Soon, young naked figures appeared on most of the balconies and terraces of the house and began jeering at the owner of the yacht, making fun of his protests. Unfortunately, he turned out to be the chief of the Potsdam Police, and charged my father with indecent exposure and corrupting young people's morals. The press got hold of the story and my poor father was teased mercilessly on his daily visit to the stock exchange. When his case was heard, he was found guilty only of indecent exposure, and was fined one mark.

Louis next to the family painting by Joseph Oppenheimer. The Hagen children and parents are in front of the Villa Hagen. Louis is the child leaning on his elbow.

Villa Hagen, where Louis grew up. Bertinistrasse, Potsdam, was commissioned by his father.

View into the sports room/gymnasium at Villa Hagen.

View of the landing, looking out the garden window.

4

Schloss Lichtenberg

As the Nazis came to power, our happy, close-knit family life came under threat. I had always known about my family's Jewish background but had never thought much about it. We were like all the other families in Potsdam: German. Our family had lived in the country for hundreds of years and my father had been awarded the Iron Cross (Second Class) for his service as a German naval officer. We never knew this until after his death, when I found it tucked away among his less important bits and pieces.

Gradually, I was made more aware of my Jewishness, at first through antisemitic propaganda in the newspapers and on the radio. Our friends remained our friends and our acquaintances assumed we were fellow members of the master race – we did not look particularly Jewish, and certainly nothing like the crude cartoons in *Der Stürmer* and other Nazi publications – but it was not long before I began to feel like an outsider. I was expelled from the rowing club, excluded from dances and public functions and was made to feel like a criminal if I tried to take out an Aryan girl. I was constantly reminded that, in the opinion of the new German political system, I was a second-class citizen; but even then I felt, as my father did, that this squalid madness could not possibly last.

It did last, and in May 1934 I was arrested by the Brownshirts (SA) and sent to the concentration camp of Schloss Lichtenberg near Torgau.

The year before, I had written a postcard to my sister Carla: 'Toilet paper is now forbidden, so there are even more Brownshirts.' This pathetic schoolboy joke had been intercepted by a maid. When my mother later caught her stealing jewellery and sacked her, she triumphantly showed my mother the postcard, which she had kept for just such an occasion. She told my parents she would denounce me to her boyfriend in the SS if they did not reinstate her.

It was an agonising decision for them. They talked it over at length, and in the end, decided they must not allow themselves to be blackmailed. Of course, they could not have imagined what the consequences would be for me. Who would believe a teenage boy could be arrested and taken to a concentration camp for a childish joke?

But one beautiful summer morning, two policemen came to the BMW factory at Eisenach, where I was working as an apprentice, and arrested me. At the police station, they emptied my pockets, took down my personal details, and locked me up for the night. They were vague about the reason for my arrest; a uniformed Nazi storm trooper who came to interrogate me next day told me only that it was 'political'.

For the next few days, I was shunted around by train to different small-town police stations, where I was endlessly interrogated by obviously amateur, pompous Nazi Party members who took copious notes. The interrogations were unpleasant, with verbal abuse, threats, and spotlights glaring down at me, but I was never physically hurt. I feel sure the reason for this was that the sessions took place in police stations, and the police at the time still believed in protecting the rights of the individual – until found guilty in a court of law. Neither I nor any of my family had ever belonged to a political party or attended a political rally; my only crime was that I was 100 per cent 'non-Aryan'.

As my journey continued, I was joined by more prisoners until there were about twenty of us: Jews, dissidents, Catholics, communists, and pacifists. We talked very little among ourselves because we were all afraid of informers.

Eventually, we arrived at our final destination, Schloss Lichtenberg, less than an hour's drive from my home in Potsdam. The Schloss was not so much a castle as a fortified, rather dilapidated farm, newly painted bright yellow. Fifty of us were housed in the large grain and haylofts, sleeping on rough palliasses on the floor. There was one water tap and a basin, which the older men used to pee in when they could not wait.

In the middle of the night, the younger prisoners were often woken by drunken guards and taken to their mess. There, to howls of laughter, we were ordered to pull down our trousers and pants and get on all fours. Then they took turns to slap our naked bottoms, either by hand or with little riding whips, which they kept tucked away in their jackboots. I had the impression I was often singled out for special attention because of my youth and class. As they beat and abused me, they called me a spoilt Yiddish brat crying for his fat Yiddish mama. They would teach me a lesson; they would show me who was master in Germany today.

They made fun of my uncircumcised penis. 'How typical of your race,' they jeered, 'to go to such lengths to deceive us.'

More often than not, these sessions happened over the weekend. It was their idea of a bit of fun to round off a good night out.

Some of them obviously felt sorry for us and tried, without much success, to restrain the others when they began to get frenzied and draw blood. There was one very blond and good-looking youth, not much older than myself, who tried to make his mates go easy on me by pretending to join in the spirit of the occasion. 'That's enough. Let the poor bastard go. Can't you see his Jewish arsehole is already bleeding? We don't want it to get infected.'

Sometimes he managed to slip me out and lead me back to my quarters, putting an arm around my shoulders and mumbling drunkenly, 'Don't mind them. They're good chaps at heart, just a little drunk.' I had a hard time not leaning on his shoulder and pouring out my misery and frustration. But I always blinked back the tears; I dared not show weakness or gratitude. He was a degree or so better than the others, but he wore the same brown uniform and jackboots. When he left, I fell back on my sack of straw and cried uncontrollably. I was comforted by a fat, middle-aged communist called Wolfgang, who became my friend.

One day, while I was playing chess with Wolfgang, I noticed a group of men crowding around the window facing the courtyard. We got up to see what they were looking at, but before I could get there, Wolfgang pulled me back. 'There's nothing there, Büdi,' he said. 'Let's get on with the game.' But I insisted on seeing what was happening. If ever I got out, I wanted to be able to tell people what was going on. It turned out to be the cruellest, most shocking thing I have ever seen.

It was a very hot sunny day. A group of SA men in their shirtsleeves were standing round the farmyard pond, where there were usually a few ducks swimming and a couple of pigs cooling themselves in the mud. That day there were neither ducks nor pigs in the pond, but instead, four prisoners splashing around, entirely covered in mud, moving as if in slow motion because of its dragging weight. They were trying to crawl out of the pond, but whenever they reached the edge, the SA men kicked them back in, laughing and shouting. I could not go on watching and turned away. Later, I learnt that none of them survived.

When the guards could think of nothing better to do, they gave each of us a bucket of water, then chased us around the courtyard. If we spilt any, they beat us. I was young and very fit, and I managed well enough, but some of the older men were soon exhausted in the blistering heat and collapsed. They were then kicked and beaten until they got up and started running again. The weakest were chased into the muddy pond to 'cool down' before, covered in mud, they had to start running again.

Perhaps because I was the youngest and the son of a well-known banker, I was chosen for the duty of emptying the latrines. They were made from thick wooden planks, about 4m long and 1m wide, and very heavy, so that four people could only just lift one when it was full. We had no spades or tools, so had to empty them with our bare hands. The iron handles of the boxes cut into our palms, which soon became infected; the pain increased each time we were made to do it.

One day, as we were standing miserably to attention in the castle's courtyard, the gates opened to let in a big black Mercedes flying the Nazi flag. I knew the smartly uniformed man who stepped out; he was Judge Engelman, the father of one of my school friends and a senior civil servant in Potsdam. I heard him say he had come for me, but I did not dare move. I was taken to the commandant in the guard-house, who told me that if I ever revealed to anyone what went on inside the camp, they would 'get me', wherever I was, and then I would never 'get out'. He handed me a statement to sign, acknowledging that I had been well treated and that all my possessions had been returned to me. I signed it.

Once in the Mercedes, the judge turned to me. 'Now, what happened?' he asked. I wanted to tell him everything, but I hesitated. I was afraid the two SA men in the front of the car could hear through the glass partition, and I hardly knew my rescuer. I had met him only once at a school function, and he was a prominent member of the Nazi Party, having joined when it still seemed to many conservatives that it represented new hope for Germany. But I remembered those still in prison and knew that I had to take the risk, so I showed him my hands. I don't think I would have dared to do that today, now we know the full extent of the horrors that were committed by the Nazis.

It was time for me to leave Germany. My older brother and sister had already left, KV for New York and Nina for Paris, and Herr Popp (*Generaldirektor* (managing director)) had been obliged to sack me from the BMW factory on the grounds that I was 'non-Aryan'. It was becoming difficult to find countries willing to accept German refugees, but an English banker friend, Sir Andrew McFadyean, managed to get me a British permit and a job with the Pressed Steel Company in Oxford.

My mother fitted me out with a wardrobe of clothes she thought suitable for England, and my father gave me a list of names and addresses of old friends who could be helpful to me. I left my family with sadness, but my country with great relief. I knew Mumpitz would be waiting to meet me in London.

Schloss Lichtenburg
in Prettin, Südseite.
(Global Fish via
WikimediaCommons,
CCA-SA 4.0)

Louis as a young man.

5

Making Friends

After several weeks in Mumpitz's flat, I looked at the list my father had given me. I rang the first name on it and found myself invited for the weekend to a country house, a large, pillared mansion that was certainly impressive, but no more so than many of our friends' houses in Germany. What did surprise me was the formality of the daily routine. Every evening the party changed into dinner jackets and was served by a butler and two parlour maids. The elaborate courtesy and quiet conversation was pure P.G. Wodehouse (our English tutor had been fond of the author and used to read us his books) and impressed me no end.

After dinner on the first evening, the men gathered around the log fire while the women 'retired'. I became a little tipsy, unaccustomed to the quantity of port and brandy drunk, and, having been asked, held forth about the evils of the Nazis and how the only way to stop them overrunning Europe was with military force. 'Diplomacy, rational argument, negotiations and compromises,' I said, 'would only be ridiculed as signs of weakness.'

When I finished, there was silence, and embarrassed eyes looked everywhere but at me. At last, my host spoke. He realised, he said quietly, that I felt bitter at having been thrown unjustly into a German prison, and at having some of my activities restricted, but Germany had considerable problems and should be given the opportunity to deal with them without outside interference. There was such a thing as loyalty, and it wasn't playing the game for a young man to turn against his own country – to come over here and make anti-German propaganda. England had given me refuge and it wasn't right for someone in my position to tell the English how to run their foreign policy, especially when my views could be understood as warmongering.

The other guests seemed to agree but were ready to forgive me on account of my youth and unfortunate position. They went out of their way to be kind to me for the rest of the weekend.

A month later, I was invited with Mumpitz and Kay to stay with the Bevans. They lived near Reading in a small cottage with low ceilings and wooden beams. Nye was a big man in every way, bubbling over with ideas and a vitality, confidence and friendliness that were immensely attractive. He loved the good things in life, but never forgot his mining background in Wales or those he had shared it with. Jennie Lee, his Scottish wife, was also a Labour MP, and a beautiful, intelligent, idealistic and passionate woman. Her father had been a miner, too, and her parents lived with the couple, looking after the cottage and garden while Jennie and Nye were busy with their political life in London.

After dinner, and several bottles of good wine, we settled down to talk. They asked me all about Germany and how I liked being in England. I had learnt my lesson at the country house party and avoided any criticism of the British Government for its failure to oppose German expansionism. Instead, I praised, sincerely, the British sense of fair play, the success of their democratic institutions, their tolerance and sense of humour.

Jennie rudely interrupted me. 'You don't know anything about it. That's the trouble with you bloody foreigners. You tell us how wonderful everything is here when you haven't seen anything except a few overfed Londoners. You should go with Nye down to the Rhondda Valley and see the inhuman conditions the miners live and work in, or come with me to Scotland and see the endless queues of unemployed waiting for a bowl of soup from the local welfare organisations. You'd soon stop telling the British how wonderful they are.'

I sat in numbed silence. Then Nye banged me on the shoulder. 'There, there, boyo, you weren't to know. You've only been here five minutes, isn't it?'

Jennie was immediately smiling and apologetic, but it was a long time before I dared either praise or criticise my new country again.

It was now early March 1936, time for me to leave Fitzroy Square and present myself at the Pressed Steel Company in Cowley. I was told Cowley was part of Oxford and an easy bus ride from the train station, but I could see no sign of a bus stop and none of the passersby I asked seemed to understand this foreigner with his limited command of English. Finally, a tall young man noticed my difficulty and offered to help, warmly inviting me to come with him to his 'digs', a shared flat in Wellington Square.

My new friend was tall and good looking, with curly black hair and dark eyebrows that met in the middle. He was obviously intelligent, seemed to find life amusing and didn't pay much attention to how he dressed. He was a postgraduate student in biology called Peter Medawar. When we reached the flat, Peter sat me down on a worn-out sofa and assured me we could sort everything out over a cup of tea – the standard British response to any problem. Presently, other students arrived, and again I found myself answering questions about Germany, until a very attractive, slim, dark-haired girl came in, and immediately all that was forgotten.

I watched, fascinated, as the others crowded around her, all talking at the same time. Jean Taylor was a science student in her first year at Somerville College; she and Peter were the best-looking couple I had seen in England and would later marry.

When Jean heard my story, she insisted they drive me to Cowley, but by the time we got there the factory had closed, and besides the guard at the gate had no idea who I was.

Peter drove me back to Wellington Square and I spent the night there. I reported to the factory the next morning and was started on my first job, on a conveyor belt.

6

On and Off the Factory Floor

The Pressed Steel Company's principal business was the manufacture of motor car bodies, mainly for Austin and Morris. My work on the conveyor belt was to smooth off welding ridges with a large electric sander. It was hard, monotonous work but I didn't care – anything was better than life under the Nazis in Germany.

I earned £3 a week, the standard wage for young workers. Half of it went on board and lodging in a bedsitter where my breakfast and dinner were served to me in my room by the landlady. I spent the rest in the factory canteen, on beer, going to the cinema and, much too rarely, on taking a girl out for the evening. It wasn't an affluent life – I was always broke two days before payday – but I managed somehow.

Although it was forbidden to send money out of Germany, I still got help from home. Every week, I sent a parcel of dirty laundry to my parents in Potsdam, which was returned ten days later with everything meticulously washed, ironed and mended. In the pockets, sleeves or trouser legs, I usually found hidden a 10 mark note, and occasionally one for 50 marks, along with a box of chocolates or a sausage. It was great getting practical help, and the almost physical contact with my mother was enormously comforting.

I boarded with a nice working-class family near the factory. They were good people and tried to make me feel at home, but the trouble they took in helping me with my English caused problems: I soon realised they spoke very differently from my university friends. At times it was almost as though they were speaking another language, and this class difference disturbed me.

My upbringing had been based on the idea of equality between people. Although at home we'd had maids, a cook, a janitor and gardeners, we were always taught that we were all equal. We children addressed the servants as Herr, Frau or Fräulein, and they didn't speak any differently from us. However, in England, I found that matching my pronunciation to that of the university people with whom I socialised, distanced me from my landlady, her family and my workmates at the factory. This made me feel uncomfortable. My friendship with the Medawars lifted me a step or two up the British class ladder and widened my horizons, but at the same time it showed me what a gulf there was in England between those who had a higher education and those who left school at 14. I seemed to be the only person in the whole factory who had any social contact with university people.

I was also the only foreigner out of five thousand or so workers, and was deluged with questions and comments. They laughed a lot over my broken English and teased me good-naturedly. They once assured me that the polite form of greeting was, 'Good morning. Had it in lately?' When the foreman came on to the shop floor and I proudly repeated what I had learnt, they nearly made themselves ill laughing.

Among the students, too, my background was of interest. I was amazed at how curious – and how ignorant – they were about what was happening in Germany, particularly how much of the Nazi propaganda they believed. Some really thought Hitler had built the *autobahns* to create work for the millions of unemployed, rather than as part of his grandiose plans for the creation of a greater German Reich. They accepted the official German line that the expansion of the army and navy, and the general rearmament programme, were necessary for the maintenance of law and order, and to restore the nation's self-confidence and pride. Worst of all, some had swallowed Goebbels's myth about the 'Aryan Race'. I was told more than once that the Germans and English were of the same 'Nordic stock', that we were like brothers and had more in common than any other two nations. Having observed both from very close quarters, I found this idea ridiculous.

One evening after leaving the factory, I went into one of the nearby pubs for my now usual pint of bitter. Above the hubbub, roars of laughter were erupting from a very noisy group in the corner. One of them was from the factory, and he called me over to join them. He was a large guy of about 30 with a protuberant beer belly he referred to proudly as his 'brewer's goitre'. He was always ready with a joke and a laugh and had a roving eye for any passing and passable female. His name was Bill Brandon; he had studied maths at Oxford, was a brilliant engineer and had become head of the research laboratory at Prestcold Refrigerators, a subsidiary of Pressed Steel. That evening we got on well, and a week or so later I was delighted to learn that he had applied for me to be transferred to his department.

Making refrigerators was a new venture for Pressed Steel, and the machines were of poor quality. Bill Brandon, now with my help, was supposed to improve the existing models while developing a new model at the same time. I knew nothing about refrigerators, nor, in spite of my almost three years' apprenticeship with BMW, was I a trained engineer. Bill, however, was skilful in covering up for me.

One of the most obvious faults of the Prestcold fridges was vibration. They shook like a clapped-out Baby Austin climbing a steep hill in second gear. Anything set on top would crash to the ground within minutes of the machine being switched on. I thought the problem might lie in the rubber washers on which the motor was mounted, so I experimented with those. When I tried thinner ones, the vibration remained the same but the noise was deafening. When I put in thicker ones, there was a breakthrough: movement was transmitted to the legs of the cabinet, which began a slow, crabwise movement across the floor. This attracted the attention (and jealousy) of the other engineers and they began to try out my modification to see who could make their fridge move the fastest. Soon we were running fridge races down the length of the lab.

Bill Brandon, showing his real talent as a team leader, opened a book on the races, selling bets in other parts of the factory and laying down handicap rules based on horsepower and weight. Still mindful of the purpose of the lab, he limited the racing to half an hour after lunch, two days a week, and insisted we work harder at our real tasks to make up for lost time. He himself worked steadily away on the new model and a few months later produced a prototype that was almost silent and, rather disappointingly, remained in the same place.

I had, at this time, all a young man in a foreign country could want, except for one thing: sex. It was the old problem of never the time or place or loved one all together. Well, I had the time and a number of potential loved ones, but I had no place. Back home in Potsdam there were so many bedrooms, outhouses, nooks and crannies that this had never been a problem. But here in Oxford, my landlady would have been horrified if I had tried to take a girl to my room. The female undergraduates were, in those days, strictly chaperoned, and the girls from the factory lived either at home or in bedsits like mine with similarly sharp-eyed landladies.

One weekend, I was invited to a lovely Queen Anne country house belonging to the lord lieutenant of the county. It was terribly civilised and English. Not a large party, just my friend Charles from New College, his 15-year-old little sister Grace, plus a few interesting neighbours. The talk was animated and well informed, mostly about international affairs and especially about Germany. The guests left at around half-past nine and, punctually at ten o'clock, my hosts announced that they were retiring.

I did not feel like going to bed yet; I was too stimulated by the talk and the company. I would have liked to stay up talking with Charles and his sister, but Charles yawned widely as he rose to wish his parents goodnight, and soon I was shown to my room in a distant wing of the rambling old house. There was nothing to do but climb into bed, think over the evening's events and reach out for one of the books that were piled on the bedside table.

Half an hour later, I was just beginning to feel drowsy when I heard a creaking noise on the stairs. The sound came along the passage, then stopped. The door groaned open and a small figure appeared. It was Grace, dressed in a white flannel nightgown. She tiptoed into the room, carefully closed the door behind her and came to sit on the bed. She had seen the light under my door, she said, and since she wasn't tired, thought she would come in and talk. I was very concerned that someone might discover her in my room, but she was quite unconcerned and chatted away about the guests that evening and the conversation.

After a while, she said she was cold. When I suggested it was time for her to return to her room, she shook her head and crawled under the covers.

Now I was terrified, and my instinctive response to an attractive female began to feel very uncomfortable. I gave Grace a reassuring, avuncular smile and made great play of moving over to one side of the bed.

'What are you doing, Büdi?' Grace asked with a smile.

'Just making room for you and getting comfortable,' I answered truthfully.

'You're not comfortable, Büdi? Is this better?' she asked in a way I could not misunderstand.

Later, I asked if she had ever had sex before, and she answered, 'Oh yes, whenever my brother brings a friend home. You've seen for yourself how boring it is here. And one has to be nice to one's guests.'

A year or two later, when I had a flat of my own in London, I found no difficulty in being nice to her when our positions were reversed.

Louis when he was working at the BMW factory, Eisenach, 1933.

7

A Salesman in London

One day, my easy life in the engineering lab was disturbed by a message summoning me to a meeting with the general manager, Hank Müller. I felt as a devout Catholic might on being called to an interview with the Pope. I wondered what he could want from one of his lowest-paid and youngest underlings. I wasn't exactly scared; I knew I had done nothing seriously wrong, but I also knew I had hardly overexerted myself on behalf of the Pressed Steel Company.

When I entered the manager's office, I was greeted by a stocky, grey-haired American with sharp blue eyes who seemed curiously ill at ease. He asked me to sit down and offered me a cigarette. I had a feeling of déjà vu, remembering my meeting with BMW's Herr Popp.

Mr Müller cleared his throat and stared down at the desk. 'This is one of the most awkward things I have ever had to do, Mr Hagen.' He looked up at me. 'I feel really bad about this, but I have to tell you that, with immediate effect, you are dismissed.'

I took a deep breath. It was quiet in the room. Mr Müller stood and walked over to the window. 'You see, the political situation with regard to Germany, as you will know better than anyone, is now critical. And for us the number one priority is rearmament. We've just received some armament contracts from the War Office, and unfortunately, one of the stipulations is that we may not now employ any foreigners.'

I was shocked. What was I to do? I had a work permit only for this job, no savings and no family to fall back on.

Mr Müller could see my consternation. He asked what I would do and whether my family were in England. When I explained my situation, he looked thoughtful, then said that, for the moment, I would remain on full pay. Meanwhile, I had better pack up my things and go home while he tried to work something out. He would be in touch in a few days.

A week later, I was once again summoned. This time Mr Müller seemed more relaxed. 'Well, Mr Hagen, I think we've found the answer to your, and our, problem. If you agree, we could transfer you to the sales and service department in London. As this is a completely independent organisation, the War Office has agreed that this would not go against their requirements.'

I thanked him warmly; I could not have been happier. Although I would miss Peter and Jean, the prospect of a new life in London was very exciting, and I could come back at weekends and visit all my Oxford friends.

The next Monday, I showed up at the very stylish sales office of Prestcold Refrigerators in Albemarle Street near Berkeley Square in London's West End. The receptionist looked like a mannequin, and beautifully dressed secretaries and elegant young men sailed busily about with clipboards, folders and important-looking expressions. I felt out of place and the bored-looking receptionist clearly thought the same about this scruffy young German. But I'd been sent on the express order of the managing director, so she telephoned for the general manager's secretary to come and get me.

Soon I was sitting at my own desk, in my own office, with the use of a secretary who, luckily for my concentration, was about my mother's age. She fussed over me for the whole of my very short career as customer liaison officer specialising in technical queries.

I spent some time exploring the neighbouring offices to find out what their occupants did. I was sociable, and people enjoyed talking to me about my life in Oxford and Germany. I got on well with the pretty secretaries, who took pity on me when they found out I was all alone in a strange country.

Every so often, I would go back to my own office and try to deal with the growing pile of complaints from unhappy customers. But my colleagues soon found out that I knew nothing about sales charts, targets, discounts or percentages, and was unable to write or dictate letters. In fact, I was scarcely able to read the letters that I received. Finally, the reality of my working situation became apparent even to the general manager, who interviewed me again and told me he had come up with a position for me that might be more useful to the organisation.

In the Oxford factory, my technical knowledge of refrigerators had looked pretty pathetic, but here among the elegant salesmen, I ranked as an expert. So, I was released from having to answer customers' complaints and sent instead to be a 'specially qualified consultant' to the more troublesome and unlucky customers whose machines had repeatedly gone wrong. My job was to calm them down and make some sort of show of finding a solution to the problem. I was to be not just another mechanic trying to botch up their faulty machinery, but a well-spoken foreign engineer who knew about these things and who, even if he couldn't make repairs on the spot, would pass on technical information to the lab. I used my carefully studied Oxford vowels, and the act normally worked quite well. But all I could really do was install a new freezer unit in the hopes that it would last a little longer than the original. All too often it didn't, and the second time around the customers weren't quite so impressed by the smooth-talking foreigner.

I telephoned Bill Brandon in Oxford and begged him to get the new refrigeration unit into production before I got lynched. I even made a practical suggestion: that he might modify the new unit so it would fit into the old cabinet. That would be cheaper than exchanging the entire refrigerator and may help to retain some rapidly shrinking customer goodwill.

Bill managed to complete the modification in an extremely short time, and within a few weeks, the new unit started rolling off the production line. From that moment, my working life was transformed. I could now visit all the unhappy customers and persuade them to change their compressor unit for one of the new design, with a two-year guarantee and a quite modest payment. With a £2 bonus for each exchange unit I sold, my daily round through the suburbs of London became quite enjoyable.

Not only did I make more money, but some of the revived goodwill rubbed off on me personally. Lonely suburban housewives would offer me tea and sometimes more. My employers were very pleased at my volume of sales and never realised how much overtime I was putting in. One of my grateful new customers was Peggy, a jolly young widow in Fulham, who would from time to time phone me at the office to request a 'service call', which I always answered promptly. On these occasions I would fill in my worksheet, 'Equipment inspected and left in satisfactory running order'.

The assistant accountant, Herbert Anstiss, a tall, gangling, serious-minded young man, noticed these repeated calls and asked me about them. I didn't want to lie, so I told him the truth. Instead of ticking me off, he wanted to know more. Although shocked, he was impressed by my successes – almost as impressed as he was when he learnt that my father was a banker. His snobbery turned out to be very useful.

From then on, he was always ready to be my banker. Although I was by this time earning a lot more money, I was spending even more, and the three or four days before payday were invariably difficult. With Herbert's help, I was always able to get an advance.

With more money to spend, a car on expenses, and my libido well satisfied, it was time to find a decent place to live. I was tired of grubby bedsitters and landladies who objected to female company after nine o'clock in the evening. I looked at many one-room flats, but they were all expensive and poky, and I soon realised that the larger ones were relatively cheap. I found a lovely four-room flat with a garden near the river in Chiswick, and had the idea of asking Herbert if he would like to share it with me. To have him under the same roof as me might be a way of consolidating my line of credit. Herbert agreed with alacrity. I suspected he might also be suffering from landlady trouble. Although he looked a paragon of respectability in his sober blue suit and old-fashioned collars, his fiancée, Doreen, was a vivacious girl with prominent teeth, an ample bosom and a sexy look in her eye, not only when she was gazing at Herbert.

We moved in, got the place looking tidy and promptly gave a flat-warming party to which I invited Peggy. The evening went well and, apart from our own friends, we were joined by several other tenants of the house who came to complain about the noise, then stayed to add to it. Towards midnight, the numbers were thinning out and Herbert began to look nervous. He had several times reminded Doreen about the time of the last tube. She hadn't shown much concern and now it was too late. There were just the four of us left in Herbert's room.

I took Peggy by the hand. 'I think we'll leave you two to it,' I said. 'Goodnight. Doreen. Goodnight, Herbert.'

Herbert didn't speak but his very prominent Adam's apple worked up and down like a yo-yo.

'Nighty-night,' said Doreen confidently. I caught her eye as I closed the door. She gave me a very clear slow wink.

The next afternoon, Herbert and I cleared away the remains of the party and settled down to a much-needed cup of tea. 'Did you have a good night?' I asked him.

Herbert put down his cup and looked intent. 'It was marvellous, Büdi, marvellous.'

'You could always ask Doreen to move in with you,' I suggested. 'I would be very discreet, and of course, I'm fairly well occupied myself most of the time.'

Herbert looked decidedly uncomfortable. 'We couldn't do that,' he said. 'Maybe on special occasions, like Christmas or Easter, but we really don't want to make a habit of it.' And they never did.

My social life was expanding. Friends from Oxford introduced me to their friends in London, several of whom worked or hoped to work in the theatre, and through them I got to know some of the fringe theatres in and around London. Here young playwrights, struggling actors and tyro producers got valuable experience before moving on to the West End. I had a car and fairly flexible working hours, and as I was the sort of chap who could build things and knew something about electricity,

I was always welcomed as stagehand-cum-electrician. No one was paid, of course. On the contrary, during the frequent financial crises, we were all expected to chip in whatever we could afford.

The theatre I liked best was the Barn Theatre, just a three-minute walk from the pretty village of Shere, near Guildford in Surrey. The stage and auditorium, holding an audience of about a hundred, was in a wooden barn that must have been built in the 1800s. The surrounding outhouses, byres and stables served as kitchen, green room, wardrobe and sleeping accommodation for everyone involved in the current production. It was surrounded by lovely open country, most of it, including the theatre site, owned by the local squire, Reginald Bray, known as 'Uncle Reggie' to everyone in the theatre.

He and his sister, 'Auntie Sylvia', lived in the old manor house and were the sponsors and driving force behind the theatre. It was open only in the summer, and I spent as much time there as I could, pitching my tent for the whole season. There were, of course, plenty of aspiring young actresses with whom to fall freshly in love, and I spent many happy nights lying out in the meadows or snuggled in a pile of hay in one of the barns.

When the weather was good, the place was idyllic. We ate outside on long trestle tables and everybody helped with the domestic chores. Groups would rehearse or learn their lines in the shade of the oak trees, in the meadow or on the bank of a little brook that ran close by. Others would be busy building or painting sets, or just lazing about, talking. At night we sat around a big bonfire and entertained each other, playing musical instruments, performing sketches or parodying current plays and films.

I don't know if any of the musicians achieved fame later, but some of the young actors certainly did, among them Michael Gough and Peter Ustinov. John Burcell was one of the most successful producers. Although he had been crippled by polio and needing crutches to walk, he had immensely strong arms, and if anything needed to be done in the flies – the cavernous upper reaches of the stage – he climbed up using only his hands and swung himself from rope to rope as if he were a great ape at home in the forest canopy. He went straight from Shere to be one of the administrators of the newly formed National Theatre Company.

The plays we put on ran for a fortnight, and on the last night Uncle Reggie always hosted a party around the campfire. Crowds came down from London. It was, for me, a golden time of freedom and comradeship.

It was on the stage of the Barn Theatre, one fine Sunday morning, that I heard Neville Chamberlain announce the declaration of war between Great Britain and Germany. Somebody turned the radio up to full volume, drowning out the argument on stage between director and playwright. Everybody fell silent. When Chamberlain had finished speaking, the silence went on.

I felt dumbfounded, unable to take in what had happened. I had always advocated military force as the only way of stopping Hitler, yet the thought of war appalled me. I knew it would affect the lives of millions of people and change the whole structure of Europe. Nothing would be the same again. As for me, I was now technically at war with my host country.

8

The Phoney War

When the Pressed Steel Company was ordered to stop producing refrigerators for the duration of the war, I was once again sacked. I received six months' pay in lieu of notice, only three months of which had already been advanced by Herbert. We had to give up the flat in Chiswick, but I kept my car and lived a nomadic life, moving between the homes of understanding friends.

During the whole of this period of the so-called 'Phoney War' before the fighting began, my happy-go-lucky life left me little to worry about. I often went up to Oxford to stay with the Medawars and their daughter, Caroline, who was also my godchild, and my efforts to find love continued unabated.

I worried about was what would happen to me as a German in an enemy country. So, when a good friend from Oxford, Peter Gatsby, asked if I would join him in running a pig farm not far from London, I thought this an excellent idea. I was to be his assistant in return for board and lodging, and we would share any profits. He soon had a contract drawn up by a solicitor, but on my way to sign it, I passed a recruiting office set up in a new corrugated iron Nissen hut.

I had taken it for granted that, as an enemy alien, I would not be eligible to join the armed forces, but without thinking, I turned back and went in. I was immediately grabbed by a hearty, back-slapping sergeant, who plonked me down on a chair in front of a trestle table and told me to fill out an application form. When I said, rather diffidently, that I was an enemy alien, he didn't seem unduly surprised, but told me to sign the form and leave the rest to the army. Then he shook my hand vigorously and congratulated me on becoming a member of His Majesty's Forces. When I asked, 'What now?' all he said was, 'You will hear from us in due course.'

I stood outside in the road, shocked at what I had just done. But the more my head cleared, the happier I felt. How could I, with my Jewish ancestry, having preached the use of force as the only possible solution to the Nazis, sit out the war in the countryside?

While I waited for my call-up papers, I continued my bohemian life, staying with friends here and there for a few nights at a time and working as a stagehand and odd-job man. I didn't worry about the papers, as many friends who had volunteered had not had theirs yet either.

One day Sir Andrew McFadyean, with some difficulty, traced my whereabouts. He told me he had an important message from the Home Office: I had to appear in front of a tribunal that would determine whether I was to be interned on the Isle of Man, in Canada or elsewhere, or was to be allowed to remain in the UK. I thought this could only be a formality. I was a Jew, I had been in a German concentration camp, and I had signed up for the British Army – how could anyone possibly suppose me to be a Nazi sympathiser? Sir Andrew offered to come along and support me, but I told him I didn't think it would be necessary.

I reported to the town hall in which the tribunal was being held and was shown into a grand room where, behind an ornately carved table, sat two nondescript civil servants and a rather pompous-looking chairman. The moment I came into the room, he barked out, 'How did you get here?'

I thought this a strange question, but told him I'd come in my car. 'Your car. How very interesting,' he commented acidly. 'And why did it take the police a whole week to contact you?'

'Perhaps because I was away in the country?'

'Do you possess a radio?'

'Yes, sir, there was one in the car when I bought it.'

The chairman leapt to his feet and bellowed, 'Haven't you noticed there's a war on? As an enemy alien, you are damned lucky not to have been locked up. The least we can expect from you people is that you obey the rules and regulations. Don't you know you're supposed to register with your nearest police station, hand in your car, driving licence and radio and not move more than 5 miles from your registered address?' His voice ended on a roar and he looked pleased with his performance. I was beginning to get worried.

'No, sir, I didn't know. Nobody told me.'

At that moment, Sir Andrew suddenly appeared, like my guardian angel. He introduced himself and told the panel he was legally responsible for my presence in England. He said he had known my family for many years, would vouch for me in every respect and was quite sure that, if I said I had not been aware of the restrictions imposed on enemy aliens, then that was true. He added that I had been working in a responsible position with a leading UK company, had many British friends and acquaintances and had very little contact with other refugees or aliens. Thanks to Sir Andrew, I was saved.

The view of Germans and Austrians at the beginning of the war was very confused. No detailed record had been kept of foreigners: they might be refugees from Nazi oppression, tourists visiting friends or spies. Now all enemy aliens had to register, and most were interned, some as far away as Canada or Australia. Later, after thousands had been shipped abroad, the authorities reversed their policy and, regardless of the shortage of shipping, began to bring them back. Many lost their lives when their ships were torpedoed by German U-boats.

Anthony Michaelis, a friend who was also Jewish, was interned and sent to Australia. His father, a doctor, was not. He was frustrated at not being allowed to practise and desperately lonely without his son. When a telegram came, telling him that Anthony was missing after the internment ship the *Andorra Star* had been sunk en route to Canada, he killed himself. It turned out the authorities had made a mistake: his son hadn't been on the *Andorra Star* but on another ship sailing to Australia.

I learnt a few days later that I had been cleared by the tribunal. I was working and sleeping at the Threshold Theatre in Notting Hill Gate, helping with the sets for a play by my friend Constantine FitzGibbon.

One morning, I returned from breakfast to be told the police had been asking for me. They came back an hour later and asked me why I had not joined my regiment. When I asked which regiment and explained that I hadn't received any call-up papers, they just laughed knowingly and said, 'That's what they all say. You'd better accompany us to the station.'

At Notting Hill Police Station, I was accused of desertion because I had not joined my regiment as ordered by the War Office when they sent my call-up papers. I told the police I had never received these and that, in any case, I had volunteered and was willing to join any regiment they cared to name, but my assurances fell on deaf ears. After a supper of bread and dripping and cocoa, I was locked up in a cell for the night.

The next day, I demanded to see a copy of the papers and asked if I might ring Sir Andrew, who promised to get in touch with the War Office and clear things up. After another night in the cells, a military escort arrived to take me by train to my regiment in Devon.

Louis with friends Colin McFadyean, Marion and Dido.

9

The Pioneer Corps

My career in His Majesty's Forces began with that journey down to Exeter. I must have looked like a criminal, unshaven as I was, and dressed in the crumpled suit I had slept in at the theatre and in the cells. A smartly dressed soldier sat either side of me, wearing the bright red armbands and peaked caps of the Military Police. During the six-hour journey, they became quite friendly. They were most interested to learn that I was German and had been in a concentration camp, but they never did believe my story about the missing call-up papers.

The Pioneer Regiment's headquarters were in Bideford. Here I was quickly issued with an ill-fitting uniform and sent to the parade ground, where a roll-call was just taking place. When I heard my name, I called out loudly, 'Yes, here!' The whole company cheered and laughed; for three weeks my name had been called at every parade and now, astonishingly, there was a reply. Almost all the men in the Pioneer Corps were German or Austrian, and most were Jews or partly Jewish. As far as the Home Office was concerned, we were all 'enemy aliens'. All the officers and non-commissioned officers (NCOs) were British.

After the parade, I was greeted by an old family friend from Berlin, Georg von Wassermann. He came from a distinguished Jewish banking family; his uncle was the eminent scientist who had discovered the first, and at that time only, diagnosis for syphilis, the Wassermann Test. Georg was a tall, strikingly good-looking man in his mid-30s.

He looked gravely at me and asked me to come into his tent. There he accused me of 'letting the side down'. It was most important, he said, that we should present a united, disciplined body of men who could be trusted by the British officers. After all, we were technically enemies and the War Office would be watching very closely. We should be proud that the British authorities had allowed us to join the forces; by ignoring my call-up, I had abused their trust, and it reflected badly on all the Germans there.

Eventually, his pompous talk ceased and he put a forgiving hand on my shoulder. 'Look here, Büdi, we both come from good families. We should stick together and be an example to the others. There's a lot of riffraff here, unsavoury characters. Why don't you move into my tent, so I can look after you and show you the ropes?'

So I moved into his tent, which already housed another six pioneers, all from good upper-class families. I had to catch up on three weeks' intensive training, and Georg volunteered to be my tutor. He showed me how to polish my boots for the next morning's inspection. He did it as though it were part of some religious rite. First, the old polish had to be burnt away and the soot quickly washed off so the moisture wouldn't be absorbed. Then, he dried the boots and treated them with a special oil. Next, a thin layer of black boot polish was applied, buffed to a high gloss with a soft brush, and finally rubbed vigorously with a specially treated woollen cloth. This procedure had to be repeated several times.

Early in the morning, before reveille, Georg woke me to give me an hour of rifle drill. Since we Pioneers had no rifles, either because the British Army didn't have enough to go round or because they didn't trust us, we did this drill with broomsticks. Georg was always meticulously turned out, even at that time in the morning. He told me he had his uniform altered by a Savile Row tailor and advised me to do likewise. All the others in the tent had obviously taken his advice and my scruffy looks were an embarrassment to the group.

I found all this playing at soldiers quite ridiculous and decided to move out. Georg told me I would regret it one day, and from his point of view he was right: he became one of the first German officers in the British Army and the rest of the 'elite' in his tent were also promoted. But I could never see the sense of the spit and polish, the endless laying out of army kit, the meticulous folding of blankets, the mind-numbing inspections and the hours of marching up and down. The only benefit of it all was to make life easier for the officers, as it occupied the men for hours on end.

Apart from our daily routine of drill, inspections and learning how to march and salute, our only real duty was to guard the Devon coastline against invasion. For weapons, we used heavy pieces of wood or anything we could find, so we would not be entirely defenceless if the invasion came. This was the exceptionally hot summer of 1940, and while across the Channel the German Army overran most of Europe and the defeated British Expeditionary Force was evacuated in disarray from Dunkirk, we enemy aliens spent our free time sunbathing, swimming, exploring the beautiful North Devon coast and making friends with the locals.

This easy life came to an end when we were formed into No. 165 Company of the Pioneer Corps and posted to Cirencester, a picturesque old market town in Gloucestershire. There we were billeted in a huge, empty village hall with a corrugated iron roof and a stage at one end. We soon got to know the reason for our existence: we were to dig latrines wherever new army camps or airfields were to be constructed.

Winter was approaching and at dawn, in all weather, we were loaded into lorries and driven to one godforsaken field after another, to where the Royal Engineers (REs) had marked out holes (for the officers) and trenches (for the common soldiers). There were no motorised excavators and no tractors to help us; just spades, buckets, wheelbarrows and a heaving scrum of manpower. In spite of the hard work, we were cheerful and good-humoured. At last, we were doing something useful. And the novelty of the physical toil stimulated us – almost all the 165 Company were professional men or white-collar workers. Often one heard: '*Nehmen sie bitte den Eimer, Herr Doktor, er ist aber sehr schwer*'; '*Vielen Dank, Herr Professor, ich schaffe es schon.*' ('Please take the bucket, Doctor, but it is very heavy'; 'Many thanks, Professor, I think I can manage.')

We were a strange collection but our motto was lighthearted: 'Vy vorry? Ve vill vin ze var.' Some of the Pioneers had private means, and their wives, children, nannies and even parents moved to Cirencester to be near them.

After a few months, we had finished digging latrines in Gloucestershire and moved to Swindon, accompanied by numerous camp followers. One particularly well-off private had his mistress and horse follow him from camp to camp. Our food was first class because our meals were prepared by famous hotel chefs and a Viennese pastry cook. We had a wonderful orchestra and also a dance band, and we produced plays and cabarets. In *Journey's End* – a very English First World War play – my part was the juvenile lead, Second Lieutenant Raleigh; in Oscar Wilde's *Lady Windermere's Fan*, I reached the summit of my acting career, as the lady herself.

Louis in the Pioneer Corps.

10

Volunteering

Slowly, the War Office began to trust foreign soldiers and appoint NCOs from within our ranks. We could volunteer as drivers, batmen, cooks, mechanics, clerks, etc. to be loaned to local regiments. I had to decide whether I was going to stay an ordinary private, doing as little as possible, or play the army game of spit and polish and become an officer. Neither of these alternatives particularly appealed to me, and gradually I discovered a third choice: to volunteer for anything and everything interesting that could possibly teach me something new.

When I saw a notice on the company board asking for recruits to join the army fire brigade at Newbury, I volunteered. I was among the very few that were taken on, probably because of my training as a motor mechanic. We had to handle firefighting equipment, consisting of a dozen mobile water pumps driven by ordinary four-cylinder petrol engines. We practised pushing and pulling them into position, and connecting them to the nearest water hydrant in the shortest possible time. After a few days, we were very pleased with our efficiency. We also patrolled the petrol dump day and night, watching out for pilfering – petrol was strictly rationed for civilians. We were left very much to ourselves and duly organised our lives in such a way that each person was on duty for a whole week, day and night, and could then have a week off.

During this time, I managed to get hold of an old motorcycle, which I rebuilt and put into good working order, and I used it to go home to the Medawars in Oxford, or visit friends in the country. London was strictly out of bounds to us; it was considered too dangerous because of the air raids.

One day I was on patrol duty, standing on an iron footbridge overlooking the petrol dump, when literally out of the blue a fighter plane appeared. I was looking at it, wondering if it was a Spitfire or a Hurricane, when it dived down and opened fire. Instinctively, I hurled myself and my rifle down behind the iron sides of the bridge, and by the time I had picked myself up again, the plane was fast disappearing. By now I could clearly see on the wings the thick black crosses of the German Luftwaffe.

The alarm sirens howled. We rushed to our firefighting trailers and dragged them into position. The engines started almost at the first touch of the button, but then, one by one, they petered out, spluttering and coughing. Orders were barked, bonnets were opened and soldiers bent over their engines, probing and fiddling. I, as the expert, was called in to help, but after going through the engine drill check several times, I still didn't know what was wrong. The batteries were well charged, the spark plugs fired, the carburettors weren't flooding and the petrol gauges all pointed, as they should, to full.

Then a terrible thought struck me. I had been siphoning petrol out of the tanks for my motorcycle. By adjusting a cork ring on the thin metal rod that worked the gauge, I was able to keep my motorcycle running and the appliance always apparently full. Obviously, I hadn't been the only one pulling this trick.

There was a hell of a row afterwards, and an inquiry. Not a single culprit could be found because we all claimed complete innocence, though we did hint at having heard strange noises and glimpsing shadows when on patrol at night. The authorities had the last laugh, though: the entire company were dismissed and returned to their units.

One day, I saw a notice inviting all men with engineering or technical experience to report for an interview with the Royal Electrical and Mechanical Engineers (REME). I volunteered, was questioned about my technical experience and accepted. With some regrets, I left the secure and civilised world of the latrine diggers for an unknown future with engineers and technicians at Donnington in Shropshire. Here, ten thousand men were crammed into Nissen huts and tents on an enormous soggy field, isolated in dreary, flat countryside. It didn't seem so awful at the time because I felt proud and excited to be, at last, part of the real British Army, on the same terms as native-born Englishmen. I felt I belonged. From that day on, my background, German or Jewish, never stood in the way of my army career. I am sure that, if I'd had enough brains or ambition, I could have become a general.

One aspect of my life disappointed me greatly, though: the idiotic army rules and regulations. For example, we had to paint neat white lines around each vehicle or gun we were working on and keep all the tools, spare parts, nuts, bolts and oil rags within this area. Continual inspections by senior officers wasted more time. On these occasions, we had to lay out our newly polished tools, smarten ourselves up by washing greasy hands and faces, then crouch as if we were working on the guns.

When I developed a septic left hand, the medical officer (MO) pronounced me unfit for duty and I was sent to do clerical work in the regimental office, the place that ran the entire depot. My job was to keep up to date a huge wall chart that recorded, with coloured lights, the progress of production. It showed how many tanks, mobile guns and lorries had been prepared for shipment, and how many had been shipped out each week. All this and more was ingeniously and graphically shown in a colourful spectacle that would not have looked out of place in Selfridges' window at Christmas. Although I thought the display a bit over the top, I was very impressed at how many units were being finished each week – until I found out, from one of the clerks, that the data on the illuminated graph was a fiction, designed to impress high-ranking officers.

This trivial approach to the war was evident elsewhere, too. A lot of time was taken up with hut and rifle inspections, and again everything had to be cleaned, polished and geometrically arranged. NCOs would inspect and measure these constructions while we stood to attention, waiting to be praised or reprimanded. Once a week we were on a twenty-four-hour guard duty of two hours on, four hours off. For this, our rifles were always inspected carefully. It was a tiresome and time-consuming job to oil and polish the barrels to the high standard required, so we always kept one set of rifles in perfect condition and these were passed on from guard to guard. We hid the others under our huts, wrapped in a tarpaulin.

To break the deadly boredom, we used to answer each other's names at roll-call so that, with the help of a cunningly forged pass, we could get away for a few days. I was in charge of these 'escapes', keeping a list of the stand-ins for every man absent, the dates and times of leaving and returning and emergency telephone numbers in case we were discovered.

It was hardly worth leaving the depot for our free evenings; the pubs, cinemas and dance halls were always packed with soldiers, and as most of us had very little money, we saved ourselves for our self-organised leaves by sitting around in the NAAFI (Navy, Army and Air Force Institutes) club drinking endless cups of tea. I tried to read, but it was difficult because of the noise, the dim lighting, and the cold.

We complained about the food, the NCOs and officers, and the dreariness of life, and talked about wives, sweethearts and women in general. This was called 'subject normal', but to me most of the men's attitudes to sex did not seem normal at all. Sex was universally enjoyed, but at the same time talked about as if it were something dirty and degrading. Frequently a man would tell me about meeting a 'real scrubber' and what he had done to her against a barn door. On another occasion he would eulogise about a wonderful girl he had met and how he really wanted to get to know her. When I asked how he got on with her sexually, he was offended. He said of course he hadn't tried anything; he'd never take liberties with a 'nice' girl. When I asked how he could be physically intimate with someone he despised, but cool and distant with someone he loved and respected, he couldn't understand what I was talking about. This was how most of the men felt.

My life at Donnington changed one day after a mock battle exercise near the town of Wellington. We spent the morning crawling along hedgerows, trying to reach a small hill in the distance without being seen by 'the enemy'. At one point, we emerged into a little side road, where we were told by a red-faced second lieutenant in an umpire's sash that we were all 'dead'. Then he stopped a passing ambulance and told the driver she was 'dead and out of action', too. The driver, a cheerful woman in her 40s, suggested that in that case we could go off to the nearest pub. Her name was Minna and she was the wife of a dentist in Wellington. She told me one of the men in my company, also a German refugee, was married to one of her daughters, and invited me home to meet him next time we had a free weekend.

Minna was extremely welcoming, and so was Don, her husband. They introduced me to their youngest daughter, 17-year-old Annis. She was a foot shorter than I was and returned my gaze with very clear blue eyes. I liked her at once. She looked like a tomboy, with a lovely figure, full of energy and ready for anything. Her corn-coloured hair was cut in a businesslike bob. She grabbed me by the hand and pulled me upstairs to show me her room.

From that moment, the dreary gloom of Donnington was lit up. Again I managed to patch up an old motorcycle and organise petrol, so I was able to visit the family for a few hours almost every day. I even had my own room in their house, on the third floor, conveniently next to Annis's. We always spent most of the nights together. I assumed her parents were broad-minded and didn't object, but Annis said she was sure they saw me as a decent young man who would respect their trust and never take advantage of their daughter. To this day, I don't know which theory was right.

During this period I tried to write. Until this point I had barely even written letters, but by now I was so worked up about the waste and inefficiency at Donnington that I felt I had to do something. I poured out my complaints to anybody who would listen, including, when I was on leave in London, Sir Andrew McFadyean. He suggested I should put my criticisms on paper. I worked for weeks, setting out my complaints and suggestions for ways of improving things. I pointed out that men with technical skills and imagination were more suitable for promotion than spit-and-polish men. I suggested, too, that the men themselves should be asked how to increase efficiency and rewarded if their ideas were taken up. I knew it was essential to get them involved, so they could feel they were doing something important and doing it well. Sir Andrew liked what I had written and passed on my report to the War Office. I have no idea whether it was even read, but it gave me an outlet for my frustrations, and as Annis helped me with it and typed the final copy, it was also a very pleasant exercise.

My happy off-duty life continued. Annis and I went on bicycle tours of the district and further afield on my motorbike. One day we even reached Montgomery in Wales and made love on a bracken-covered hillside. Later that summer, when Annis had to go to hospital in Shrewsbury to have her tonsils examined by a specialist, we cycled the 12 miles. On the way back, hot and sticky from our exertions, we found a deserted bank of the River Severn, stripped off and plunged in. We were in heaven. The war could have been on another planet.

But then the autumn days grew shorter and Annis reached call-up age. She was posted to the US Army HQ at Salisbury to work as a clerk, while I was drawn to a new announcement on the company noticeboard, inviting applications for a posting to the Royal Army Ordnance Corps. I thought it would be interesting to learn something new, so I applied, was sent to the selection board and accepted.

On my way back to Donnington, I 'lost my way' and spent a few days with the Medawars in Oxford. Peter was working furiously on ways of treating serious burn injuries, work that led to the discovery for which he would in 1960 be awarded the Nobel Prize for Medicine.

At my new unit, I had scarcely unpacked when a fresh directive told us that foreigners could now volunteer for the Artillery Assault Corps. I wasn't quite sure what this was, but it sounded exciting, so a fortnight later I reported to the huge tank and artillery training grounds at Tidworth on Salisbury Plain. Here I was taught to drive and operate a tank – then learnt that these tanks were to be fitted with explosives, driven under bridges and blown up. Although I enjoyed driving the 5-ton monsters over the shell-pocked testing grounds, I didn't fancy doing the same thing with half a ton of explosives inside and enemy fire outside.

Luckily, something else turned up: aliens were now allowed to volunteer for the Commandos. I went for an interview, was accepted and soon found myself sitting in a room with a lot of keen, rather rough-looking young men being told by an even keener and rougher-looking young captain what the Commandos were for. I learnt that we were the tough guys of the infantry, specially selected for our rugged physiques, resourcefulness and cunning. We were to land at night on enemy beaches and destroy grounded planes, gun emplacements, bridges, docked U-boats, dams or reservoirs – anything of military importance behind enemy lines.

While shaving the next morning, I saw myself with fresh eyes. I flexed my muscles and felt resourceful, cunning and greatly honoured to have been accepted. I threw myself into the training with enthusiasm and felt this was something I could stick with.

But then I saw another notice: I was now eligible to join the Army Air Corps.

11

Learning to Fly

To become a pilot was the ultimate goal for thousands of soldiers of all ranks. There was an agreement between the army, navy, and air force that they would not compete, so once you joined one of them, you could never change to another. All those who dreamt of becoming a pilot and leaving the boredom of army life had only one chance. The Army Air Corps was left over from the First World War and by far the smallest regiment in the army, with only a few hundred pilots. They flew senior officers to small airfields or landing strips close to the front line, or went 'artillery spotting', reporting back to gun crews on whether they were on target. They flew very small aircraft, the types that were used privately or by flying clubs before the war.

I had never attended an interview in such a state of excitement. I felt that, if I were accepted, I would be the luckiest man alive. For the first time, I did not get lost on my way to the interview. I polished my boots and buckles, pressed my uniform, had my hair cut. When I stood in front of the selection board, I was trembling.

I meant to answer their questions slowly and precisely, in my best English. But when the chairman, a full colonel, asked me, 'Where do you come from?' I was completely knocked off balance and stood there speechless.

If I said Potsdam, would I have to explain that I was a harmless German and that I had left because of my Jewish ancestry? Would they then be prejudiced because of my Jewishness? But if I answered, 'Oxford, sir', would they then call me a liar? All these questions were racing through my head as I stood silently to attention.

'Relax, man, and answer. We're not going to bite you,' the colonel said, and smiled. His easy manner brought me back to my senses.

'What the hell,' I said to myself. 'Tell the truth and trust to luck.' So I said, 'Potsdam, Sir!'

He looked surprised, then asked me to explain. When I had finished, he said, 'My second question would have been, "Why did you volunteer?" but I see now that is superfluous.' He told me I would hear from them soon.

Two weeks later, I was queueing up for my flying gear at RAF Denham, just outside London. I was issued with sheepskin gauntlets and boots, real silk underwear, gloves and socks, a quilted overall, a leather flying helmet and goggles. A youngish, carelessly dressed officer introduced himself as Flight Lieutenant Curling, my instructor. I immediately liked his relaxed manner and the way he treated me as an equal, even though he was an RAF officer and skilled pilot, and I was just an ordinary army private. Without his patience and clever teaching methods, I might never have made it.

A rumour had gone round that unless we were ready to fly solo after seven hours' tuition we were out and would have to return to our units. This threat hung like a guillotine over all of us. We had watched so many crushed young chaps handing in their kit for good.

Our plane was the beautiful Tiger Moth. Just sitting in the open cockpit made me feel like a superman, helmet strapped under my chin, big goggles secured tightly, the motor spluttering and revving the propeller so that the rush of cold wind almost took my breath away. I had never been in a plane before and the thrill was unforgettable, mixed with fear when Curling demonstrated a loop, stall or sideslip. That first afternoon I spent getting used to the joystick and the feel of the plane's responses to the controls. It was amazing how sensitive they were – the smallest movement of the joystick, held lightly between thumb and index finger, could send the aircraft into the most complicated manoeuvres. I felt terribly ham-fisted and wondered how I would ever fly on my own, let alone take off and land.

Day after day the lessons went on, some only fifteen minutes long with three take-offs and landings, others half an hour of climbs, turns, stalls and dives. Every minute in the air was registered in the flight book, but as the hours crept up, I still could not imagine piloting all by myself. My landings were bumpy and sometimes I could not even keep the little plane flying straight and level.

When 'seven hours in the air' was recorded in my flight book, I was afraid I would soon be called to the control room and given fatal news. But nothing happened until, quite unexpectedly, one morning when my flying book showed eight hours and fifteen minutes, Curling hopped out of his seat and said, 'It's all yours. I don't dare to watch. I'm going to have a cup of tea in the canteen and try to forget about you.'

In the cockpit I stopped thinking about anything but flying. I revved the engine, taxied to the take-off position, went through the checklist – oil pressure, revs, brakes, ailerons – then gave a full throttle and released the brakes. When the Moth reached take-off speed, I pulled the joystick slowly towards me – and I was airborne.

It was a wonderful feeling to be all by myself in the endless sky, no instructor to tell me what to do. I flew the prescribed circuit and never for a moment lost the elated feeling of wonder and joy at flying alone. My landing went well – not entirely textbook, but good enough not to damage the undercarriage. Now I could start my training in earnest.

I had flown solo at RAF Denham, but for further training we were to train 20 miles away at Booker, near High Wycombe. Besides the excitement and pleasure of flying, we had lessons and aircraft maintenance every day. There was a lot to learn, but our revision was continually interrupted by kit and rifle inspections, square bashing, parades, roll-calls and other time-wasting nonsense.

One day when I was sitting on my bunk, working on the difficult navigation problems we had been given for the next day's lessons, the sergeant-major came into the hut for one of his frequent surprise inspections. I stood up, probably not very smartly, and he asked me why I was not outside drilling with the others. I said I was excused because I had to prepare myself for the next day's lessons. 'You lazy lout,' he barked. 'You've got all night to do that.'

Angrily I answered, 'I wish I had, but your stupid regulations have us in bed and lights out by ten-thirty.'

Now he roared at me, 'You're insulting a superior officer; you'll regret this, young man. Report to company office at eighteen-thirty hours!'

I was unlucky that a visiting captain, obviously a regular soldier, heard my case. The sergeant-major said, 'This man's general bearing and appearance is unmilitary and his behaviour is bolshie.' I protested but the captain made short shrift of me. 'We don't need people like you. There are thousands who would willingly take your place. Hand in your kit, get your travelling papers, and return to your unit.'

I could not believe the sudden turn of events. In desperation I plucked up all my courage, and marched back to the office. The sergeant-major barred my way as I tried to knock at the door. 'Can't you understand an order? The captain clearly said "dismissed". If I see you here again, I'll put you on a charge for disobeying an order.'

I left practically in tears, my dream of becoming a pilot shattered. After handing in my beautiful leather helmet and goggles, fur-lined boots and gloves and quilted flying suit at the quartermaster's store, I went to say goodbye to Flight Lieutenant Curling. When I told him what had happened, he said, 'What nonsense! We can't lose good chaps like you. I'll talk to the squadron leader, you just wait outside.'

Still numb with shock, I hardly dared to hope. After ten minutes that felt like hours, he came back, patted me on the shoulder, and said, 'Go to the stores and collect your gear before it's issued to someone else.' I could have kissed him. From then on I was very careful and 'yes, sir-ed' anybody and everybody.

Still, there was one critical incident before we finished training. A directive was issued, announcing that all German and Austrian personnel in front-line units were to be issued with new identity papers with different, preferably English-sounding, names, places of birth and next of kin. We were given forms to fill out and hand in within one week. There was good reasoning behind this directive: it was to diminish the chances of our origins being found out should we be taken prisoner on enemy soil. Anyone found to be German or Austrian would be shot as a deserter.

I was busy swotting for an important exam and could not decide on a suitable new name, so I missed the deadline for handing in the form. I got a vicious dressing down from the sergeant-major and was given another day to do it. When at last, just before the deadline, I had decided on a name, I found the form had disappeared. I searched frantically, but could not find it anywhere. When I confessed this to the sergeant-major, he told me, gloatingly, 'Without a new name, you are not allowed to continue the training.' At last he had found a way to get rid of me.

Again, the RAF came to my rescue and negotiated a twenty-four-hour moratorium during which I could get another form to fill in. I went home to Oxford and there Jean and Peter suggested Lewis Haig, a name I kept until a year or two after the war. The army thought it a great name because of the famous First World War field marshal, and my friends and I because of the whisky. I was registered as born in Oxford and the Medawars were appointed my next of kin.

After I finished elementary flying training, we were suddenly told we were to have six months' additional training on gliders, luckily all near Oxford. At first we felt let down, until we realised the gliders were for landing behind German lines, a tremendously exciting prospect for me. During the glider training, I received my wings as second pilot and was promoted to sergeant.

Our new training involved practising mass landings, night landings, low-approach landings, dive-approach landings and the ferrying of gliders from one field to another. Always there were rumours about when and how we would go into action. Sometimes we were called in for briefings on operations that were subsequently cancelled.

One day, at last, there was a real briefing and, towed by RAF aircraft, we took off. Hundreds and hundreds of gliders, loaded with jeeps, field artillery and fully armed combat troops, all flying towards the Netherlands. Our destination was a small town on the Rhine called Arnhem.

Group photo: Louis, standing fifth from right. Denham, 1943.

12

Arnhem

Across the Channel, I checked our maps constantly against the ground, looking for the first signs of our objective. Soon, I recognised the Lower Rhine, and a moment later spotted our landing zone: two small clearings in an area of wooded land. We made a perfect landing and leapt out to undo the heavy bolts inside the tail unit, so the jeep could be driven out. To get to our first rendezvous (RV), we had to follow a narrow sandy lane through low brushwood, small fields and single rows of trees. Everywhere we saw gliders, in the fields and even in the trees – there was a wing wedged between two big branches of an oak, a tail unit sticking right up in the air and pieces of glider scattered everywhere.

We passed a meadow with gliders parked in a more orderly fashion; obviously this was the real landing zone. We joined more and more jeeps and trailers, all filing to their various RVs. Ours was the station at Wolfheze. From there we were to find our way to a lunatic asylum. The station was tiny and its main feature was the cross-roads, one road running over the rail lines and the other parallel to the railway. Here was a terrific assembly of jeeps, trailers, light artillery and groups of parachutists. Red Cross jeeps with casualties on stretchers were passing through, and men were vainly trying to repair the railway lines. There was confusion everywhere.

All the while, we were being sniped at. Sometimes a mortar would go over and everyone seemed to disappear, but after a few seconds the confusion returned. Maps were spread out and everyone asked everyone else if they had seen or heard of their respective units.

Our glider pilot flight officers now appeared and began to collect the flight. Burdened with very heavy rucksacks, we started to move off, cursing at having to leave our jeeps behind. In single file and directed by a parachute brigade officer, we moved towards Arnhem. We were grateful for the long halts, sometimes an hour or two at a time; it was impossible to keep up with the paratroopers who wore only very small packs.

This went on until two o'clock in the morning. We were making slower and slower progress, and hearing increasing fire from the direction of Arnhem. Eventually, we got the order to turn round, and had to walk back half the way we had advanced that night. We dug in along the railway line, covering the lane along which we had advanced, to be ready for any attack that might come at dawn.

At ten o'clock that morning, fighters started passing overhead. We pointed them out to each other and soon the usual argument started.

'They are Spits.'

'Don't be an ass, anyone can see they are Typhoons.'

Although I could only tell a Spitfire from a non-Spitfire, I could see the planes were German, and before we had a chance to argue about what type of German plane they were, their machine guns started firing in our direction. We scattered and ran for shelter. Not one of our own flight was hit.

Our first job was to clear the enemy out of the wood on the hill in front of us. About thirty of us advanced. Our two officers took the centre and I was the outside man on the left flank. Not normally aggressive, I tried to work up hatred and the kind of spirit I thought was needed to assault the enemy. This was easier for me than for the others. I had only to remember the state of my hands in the concentration camp, raw and festering from cleaning out the latrines, and the Nazis' slow murder of a Roman Catholic priest in the mud of the duck pond. I was ready to take on any of those Nazis who used to strut through the streets of my hometown.

We reached the top of the hill and the fringe of the wood. Beyond the clearing in front of us, we could hear loud German voices and engines running. Bullets whizzed about us, but the thick cover in the copse concealed where they were coming from. I heard our captain shouting for the lieutenant and guessed that the centre of our advance must have suffered casualties. I was still feeling fighting mad and I called to Dodd, one of the pilots from my own flight, that I was going forward.

I got up and ran down the slope, fell behind cover, then got up again and ran on. I vaguely noticed the intense fire, but all I wanted was to get to my destination. The Germans were only 20m in front of me and I could hear them arguing with an officer who was ordering them to go forward and search for me. Instead, they threw a few hand grenades and fired off at random. I swore that if I ever got out of this idiotic position, I would never again be such a bloody fool.

But the more I listened to the Germans, the more I realised what a poor, undisciplined crowd they were. They began talking about leaving me for dead and getting their big guns into position. I threw a hand grenade, dropped my rifle and ran for it. As they spotted me, I flopped down again into the brushwood. When I tried to get up, I couldn't move my legs. I thought I'd been wounded, then found that my underpants had slipped down inside my trousers, hopelessly constricting me. I had to cut them off, with as little movement as possible, before I could dash towards the top of the hill and the fringe of the wood where our attack had begun.

I was nearly killed by our chaps, who mistook me for an enemy scout. They stopped firing only when I was near enough to shout my name and password. I told the parachute officer about the low morale of the Germans and how we could bowl them over if we attacked; but by now, they had brought up tanks and self-propelled (SP) guns, and we were forced to retire to our original positions. I linked up with my flight, who were digging in on the rear slopes of the hill. We started making tea; we had eaten nothing but chocolate biscuits since Monday morning and it was now Tuesday afternoon.

We never finished our tea because the order came to move back to Wolfheze. Five minutes later, we joined a stream of troops moving slowly back, and realised this was a retreat.

As we crossed an open field that led to the railway lines, we came under fire. By the time we reached the other side, we were no longer an organised body of men. We didn't understand this retreat and could get no clear information about what to do. By the time twenty of us had reached the Wolfheze cross-roads and met other troops lining the road, I knew we were in real danger. I was sure we should move towards the river, where the Second Army might relieve us. We took a road that was supposed to lead to the river, just east of Arnhem, but finally most of the chaps decided to return to Wolfheze. Only Dodd and I went on. We never saw any of the others again.

Just as the light was starting to fade, a string of jeeps on reconnaissance patrol came racing along the road and took us aboard. They were glad to have us as they had suffered casualties and needed to make up their numbers. We were equally glad to join them. We had nothing with us and they gave us everything. I got myself a Sten gun, plenty of ammunition and a good night's sleep in a slit-trench.

The next morning, the officer asked for a patrol to push forward to Arnhem Bridge, and Dodd and I were asked to go. We sped off again, but after ten minutes the first jeep was fired on. Everybody except the drivers got out and worked their way forward, either through the thick wood to the left of the road, or through the gardens in front of the houses on the right. We heard German voices shouting just to the right of us and saw some of the soldiers moving back towards the bridge. We kept moving steadily forward.

Then I heard the horrible crunching noise of tank tracks on the tarmac road, and not long after, the crash and thud of shells being fired into the houses. A tank advancing, firing shells, is the most frightening thing I know. We had only a little Piat gun, barely a metre long, and we could hear the German infantry advancing to screen the tank.

One of the recce men threw a smoke grenade and, under its cover, the jeeps started off again down the road, leaving only the Piat gunner, his number two, Dodd and me. We decided to try to get back to recce HQ. We withdrew through the woods parallel to the road, firing our guns to frighten the Germans, but we could hear their voices from all directions and hid in the back gardens of some houses near the cross-roads.

As Dodd and I approached one of the houses, we suddenly heard voices. Spotting a little rubbish pit hidden by shrubs about 3m from the door, we crouched down in it, just as the Germans came to search the garden. As they stood about, talking in very loud voices and giving orders that no one obeyed, the stinking, decaying garbage in the pit gradually seeped into our trousers. We could not move until, after what seemed like ten years, the Germans left the garden.

As we crawled out of the pit, in agony from pins and needles, we saw a glider pilot jump out of the back door of one house and into the next. When we joined him, he told us we must get away quickly. He had managed to shoot the whole crew of a German tank from the attic where he was hiding, and now the houses were being searched. We scrambled back through the garden to find the thicket where the two recce men were hiding, and lay there on our stomachs while mortars and shells whistled overhead.

After a couple of hours, the shelling ceased and we heard a patrol approach quietly. The lieutenant who was leading it had collected pilots from different squadrons and asked if we would like to join them. We were only too pleased to do so, following as he used his compass to lead us through the wood.

Long before we came upon the Germans, we could hear them shouting to each other. They were filing into an empty house, isolated among trees and surrounded by a wall. We fired at them and several were hit. Seeing their confusion, I asked the lieutenant to cover me while I walked forward with my Sten gun, shouting, '*Hende hoch!*' (hands up). I told them the Second Army was coming, and as they were surrounded, they had better give themselves up. But as they began filing out of the house and into the road, their officer appeared and furiously ordered them back. We withdrew.

Eventually, we reached a village called Oosterbeek, where the divisional and brigade HQ had been set up in a massive hotel, surrounded by a park and outbuildings. Here, men from different squadrons had dug extremely deep slit trenches and covered them with branches. I found a friend from my station, who offered to share his trench with me. The mortar shelling was light that night and I slept so deeply that I did not even consider the danger.

The next morning, a Thursday, at dawn, just as we were heating up our tins of breakfast, the divisional HQ came under intense mortar fire. One shell landed an arm's length from the trench I was in. There was a thunderous crash and I thought my eardrums had burst. When the barrage passed on, an officer asked for volunteers to go on patrol. I was glad of the chance to get away.

Our patrol was to reconnoitre the houses and gardens not far from our trenches and find out if they were occupied, by either their owners or German soldiers. We crept through the back gardens, crawled towards a house, listened, waited and listened again. I began to get impatient at our slow progress, so I got up and went to the next house. I knocked at the door and opened it, shouting, *'Jemand hier?'* (Anyone here?) When nobody answered, we went through the rooms to make sure they were empty, then knocked at the next house. This sped up our search considerably.

But in the next row of houses, my polite enquiry produced a stampede of German soldiers, out of the back door and into the house opposite. Before they were all inside, we lobbed grenades and hit one of them. I ran up the stairs and could see some of them crawling into a trench and taking cover behind hedges. We were not supposed to get into a serious fight, so we stayed in the house until orders came to occupy a corner house on the large road nearest to the defence perimeter of the hotel HQ.

The commanding officer, Captain Ogilvie of our party was a big young man who wore a kilt and had a very large, fair moustache. With him were five other officers and fifty glider pilots. The moment we moved in, the Germans began sniping at us from the woods and from the few houses that faced us. We could not walk from one house to another or stop at a window without someone taking a shot at us.

We set up the Bren gun to cover as much of the road as possible, and worked out a plan to occupy at least every second house until we were relieved. We barricaded the front windows, dug communication trenches between the houses, made holes in the brick walls to shoot through and set up the Piat gun by a little hole in the attic roof – the Germans never discovered its position the whole time we were there.

In spite of all this, there was not a day when a few of us were not knocked out. I remembered with disgust the weeks and months we had spent on drill and laying out boots, brushes, knives and forks for kit inspection. Why hadn't we been taught about house-to-house fighting and how to use a Piat gun?

On the second morning, we heard the sputter of a heavy Spandau machine gun from the wood opposite our house, and the whine of small armour-piercing shells that could penetrate clean through a house. This could be the preparation for a tank attack.

Lieutenant Strutt came up to the attic and offered to fire the Piat; he had once fired a practice shot, and I had once been shown how to load it. We waited until we could see the first tank clearly, at a distance of about 90m, before Strutt fired the first round. The little weapon produced a terrific explosion; he was thrown back against the wall by the recoil and covered with dust. I jumped forward to see where the shell had hit. The direction was perfect. It had fallen 20m short, but the tank had stopped, and by the time the lieutenant had picked himself up, I had reloaded.

We fired another four or five shots until the tank retired out of range. The shells the Germans had been firing had all gone past our house, hitting trees and houses further along the road. The casualties there must have been very heavy. Two of our jeeps appeared, flying large Red Cross flags, and the firing from the woods stopped as the wounded were brought out on stretchers. Whatever I felt about the Nazis, I never once heard of them deliberately firing on any Red Cross men or jeeps. They couldn't have kept more closely to the Geneva Convention. The moment the Red Cross party disappeared into our lines, the firing started up again and kept on relentlessly.

After the attack, we began to feel very hungry. We'd had no rations since we landed and most of us had lost our rucksacks. Luckily, we found all sorts of preserves in glass jars or earthenware containers hidden in the cellar. The first time I produced them, I had the utmost difficulty in persuading some of the chaps to try these 'Continental concoctions', but they were very good, and soon everyone was eating like mad. We made a fire on bricks in the middle of the kitchen floor, and before long everyone had forgotten the war and was stuffing themselves on fried chicken, pork, beefsteak and fresh vegetables, dug up at some risk from the garden. Food was collected and cooked in a big pail in lulls between heavy machine gun barrages and attempted assaults.

The noise of exploding mortar shells and bursts of machine gun and small arms fire was so continuous that after a while it seemed almost normal. We were always busy at our posts, in the trenches in front of the house, in the attics, at windows or at the loopholes we had made in the walls.

Our spirits rose after two German prisoners were brought in by the parachute major. He asked me to translate his questions to them, and we learnt a good deal to encourage us. The men, both about 40, had been in the army for only six weeks and this was their first action. They said they knew the war was lost for Germany, and when I asked why they were fighting when they felt it was useless, they said they had no choice: they would have been shot at the slightest sign of refusal to obey orders. They told us their Panzer Division had tanks, mobile guns, flamethrowers and air support. If they had known how few and vulnerable we were, we could never have held out against them.

Captain Ogilvie, who had been wounded in the arm, appeared so full of optimism and confidence that it was hard to guess how aware he was of the danger of our position. He received everyone in the officers' room – the only room still completely furnished – lying fully dressed in belt, sporran, boots and beret, on a luxurious bed covered with a quilted eiderdown. His nonchalant way of making plans with the other officers boosted our morale. Only once during the whole action did I hear any officer lose his temper, and that was when trying to wake the sergeant-major in his HQ under the best table, hidden by its overhanging lace cloth and purple velvet cover. The aggression he had shown in the barracks had deserted him here and he did not want to be noticed.

One evening, I was called from supper to the officers' room. Captain Ogilvie asked if I would undertake a patrol to discover the assembly point of the German armour that was harassing our section, and find out which houses at the end of our street the Germans were occupying.

I thought the patrol should be as small as possible, with a few men waiting on the fringe of the wood to give me covering fire if I had to withdraw in a hurry. I decided to take Sergeant Graham with me. We made our plans in the officers' room, cleaned our Sten guns, collected ammunition, put on our running shoes and tiptoed into the shadows cast by the next block of houses across the street.

Soon, we heard Germans walking through the gardens and the sound of engines running. We worked our way to a place where we could make out that troop transports were moving from left to right, to a point about a mile away. We could hear the almost continuous sound of spades digging. By moving carefully towards it, we came to a thick hedge, underneath which we found a square tunnel. Emerging from this, we saw a large open space, bounded on the far side by what looked like the outbuildings of a big country house. Two Germans were silhouetted against the buildings, digging and whispering monotonously. This had to be the German strong-point from which they sent out troops and vehicles, and to which they returned at night.

When we got safely back to the officers' room, we were touched to find that several of them were still awake, anxious about the long time we had been away.

In normal battle, our report would have been immensely valuable, enabling us to pinpoint the German position for our artillery and mortar fire. Unfortunately, we had no artillery or mortars to fire, but it was still useful to have discovered that the Germans retired from their houses to this strong-point every night. We dreamt of infiltrating those houses and giving them a surprise at dawn, but we hadn't enough men. Still, Captain Ogilvie was pleased. He sat up in bed, twirled his moustache and said, 'Well done, chaps, good show.' Graham and I collected some cushions and blankets from the floor and passed out until stand-to at dawn.

On this comparatively quiet Saturday morning, we realised for the first time that we were not the only occupants of these houses on the front line. Pale, frightened Dutch people emerged from the cellars, about ten in each, and asked timidly where they could get water, and whether it would be possible for them to come upstairs and collect blankets and food. The house next to ours had belonged to a member of the National Socialist Movement who was clearly a collaborator – there was a gold-framed photograph of him shaking hands with the local party leader, and others of German officers giving the Nazi salute, arm in arm with Dutch girls. We also found German magazines containing superbly produced weekly war reports. Reading them, it was impossible to imagine how Germany could ever lose. Captain Ogilvie cut short my study of these magazines; it was time to report on our patrol at divisional HQ.

We reached it through fire from the Germans and from our own people. The rain of mortars had increased since morning and was taking its toll on the men's nerves, mirrored in drawn faces and unsteady eyes. Not everyone had stood up to the strain. In the darkest part of the cellar I found men who had lost all their nerve and self-control; they looked like people who had been seasick for days. I knew some of them, one a sergeant who had been a close friend back at the station. He was conscientious and hard-working, and I thought had all the qualifications of a good soldier. I tried to make him come out of the cellar but he simply could not move, and there was nothing I could do. Another was a staff sergeant who had been the tough guy of our flight, always swearing and bragging and calling me Miss Hagen because I wore a nightshirt. I was astonished to see him so terrified.

We were to report to the brigadier, and found him walking about ignoring the mortars. He laid out his map and asked me to show him exactly where we had found the German hideout, and to draw him the movements of the tanks we had seen.

I began to illustrate where we had gone, using my grubby finger – apparently the one thing guaranteed to make him lose his temper. 'For crying out loud, take your filthy hand away! Get a stick and point it out properly.'

When our colonel joined the group and they began discussing higher strategy, I asked permission to retire. By now it was lunchtime, but that didn't mean much at divisional HQ, where a cup of water was a luxury. I arrived back on our street just in time for the afternoon shelling. Lieutenant Strutt and I had by now mounted the tripod of the gun on a bed and every shell we fired created a snowstorm of feathers. They made excellent plugs for my ears, which were becoming increasingly sensitive to the noise of the explosions.

That afternoon a fleet of our lumbering four-engined supply planes came in at 450m, searching for our positions. The German gunners fired at them at almost point-blank range. How those pilots could have gone into that inferno with their eyes open is beyond imagination.

By Sunday morning, the small arms fire and sniping were worse than anything we'd had before. It now came largely from the burnt-out house directly opposite, a sniper's paradise from which hand grenades were thrown against our barricaded windows. We had to replace the barricades every time the blast threw them back into the room. Were a grenade to bounce in and roll under the bed, it would explode before we could reach it and throw it back.

The afternoon was fairly quiet, but towards dusk the fire increased sharply, and before we'd realised what had happened, the Germans had crossed the road. I thought they had got into the next house, which had been evacuated to strengthen our position. I could not fire at them from where I was because they were too close, so I changed position and began lobbing grenades into the room where I'd heard noises. I was about to replenish my store when a quiet English voice called up to me, 'What do you think you're doing? Trying to kill us all?' This was the worst moment of the whole seven days and I wished myself dead.

I jumped down the stairs and ran out of the house, across to the window of the next house and climbed in. Lieutenant Strutt and three others were sitting unharmed on the floor. I said it was me who had thrown the grenades. 'Oh, it was you, was it? Thank goodness you didn't know your job. At this distance you should have waited four seconds after you pulled out the pin. We lobbed the grenades out as far as you threw them in; you're a fool not to have noticed.'

At dawn on Monday morning, the fire slowly increased from all sides. We were desperately tired. I felt like a slow, heavy machine, working at half speed. Captain Ogilvie and some other officers came back exhausted from divisional HQ and began a conference in the officers' room that lasted two hours. At once rumours began to circulate.

Finally, we were told to report to the captain for orders, in groups of four. He began the briefing in a confident voice. We were to retreat across the Rhine to join the Second Army in an orderly withdrawal, at 2200 hours, with all our arms and ammunition. He pointed out the route we should take, through woods and along little paths. The longer we tried to memorise it, the harder we realised it was going to be.

We ate as much as we could, because no one knew when we would have food again. I got my gun ready and collected rounds of ammunition. I was hoping for two hours' rest, but the firing had increased so much that we had to man our positions to give the impression of normality. We were warned not to discuss the withdrawal, for fear of listening Germans, and to move off silently in single file, fully armed, our boots muffled with sacking.

At 2000 hours, Captain Ogilvie called me into the officers' room and told me to stick by him throughout the withdrawal. We led a long, silent file of fifty glider pilots, making our way towards divisional HQ as we had done on our patrols, through the mass of slit-trenches behind it and a kind of no-man's-land towards the river, about 3.7m away.

Captain Ogilvie still seemed pretty sure of the route, but the denser the wood became, the harder it was to follow, and the more we relied on guesswork and the captain's luck. As he felt his way forward, he muttered, 'We'll make it yet, don't worry. You stick to me. Do you think we're all right, old boy?'

I hadn't the faintest idea whether we were all right, nor even where we were. Nor had anyone else, but at last we came to a farmyard we recognised from our maps and continued more hopefully on through a dark wood, turning and winding whenever we felt we were too near the enemy. I walked in front of Captain Ogilvie with my finger on the trigger of my Sten. We lost all sense of time, or distance, and simply groped forward wherever he directed us.

At last, we emerged on to a wide plain, clearly the approach to the Rhine. But as we left the shelter of the wood, what had sounded like distant firing became almost deafening as shells landed all around us. Someone came and led us to a white tape that stretched past a hedge, and along a small path running down to the river. 'We've done it!' Captain Ogilvie whispered to me as the long column followed the tape across level ground. Then he stopped in front of a dark shape. 'Oh look,' he said. 'A poor dead cow.'

As we went on, now under intense shell fire, we began to see human bodies lying all along the path, some moving, some groaning and crying for help. This was more than I could stand. One wounded man begged me not to leave him behind. We tried to lift him, but he groaned with pain and we had to lie him down again. I made out the shapes of more bodies dragging themselves towards the path, frenzied with the fear of being left behind. For the first time in the whole action, I felt panic. I dragged limp bodies towards the beach; I ran round in circles looking for someone in command and asking anyone not injured to help; I vomited and felt very faint. Finally, I was ordered to leave the wounded where they were; a doctor would be left behind to look after them.

I continued towards the river, dazed by the ghastly scenes and my own inability to help. I reached the bank and joined one of the queues of at least a hundred men waiting to be ferried over. We were told to spread out as more mortar fire was expected any minute. At last, a small rowing boat appeared and took ten men across. We began to realise how desperate our exposed position was, crouched in squelching mud, shivering and soaked with rain. I began to fear that, before we could all be ferried to safety, most of us would either be wounded or taken prisoner.

I found Captain Ogilvie and told him that, as the river didn't seem too wide here, I was going to try to swim for it, leaving the boats for those who couldn't swim. He agreed and shouted to the rest of our section that we were going on to a promontory where the river narrowed a bit. From here, we could make out the opposite bank, and the prospect of action after queueing in the mud on all fours made the captain and me feel quite cheerful. 'We'll do it again, you and me,' he said, and we took off our boots and hung them round our necks. I put my Sten gun across my shoulders, and by the time I was ready he was already in the water, swimming away from the bank.

When I went in, he was about 20m in front of me but drifting fast downstream. The current was very strong and I had to work hard against it. Gradually, swimming became harder and harder, and I struggled to catch my breath, though I was only halfway across. As I turned on to my back to rest, I thought how ridiculous it would be to drown in the Rhine after being brought up by a lake, swimming since I was 4 and after evading violent death for the past seven days.

I realised I had to get rid of everything that was weighing me down. I let myself sink while I eased the gun up over my head, then shed my battle smock and all I had stored in it: hand grenades, writing materials, fountain pen. Then my steel helmet and boots went bubbling to the bottom of the river. The difference was marvellous. I felt as I had a fortnight ago, having a picnic at Shillingford with the Medawars, when I had swum in the Thames with 6-year-old Caroline clinging to my back. I shouted and looked around for Captain Ogilvie. As there was no reply, I assumed he had already got to the far bank, so I swam on.

About 20m from land, I heard shouting from two excited figures. 'Hold on, mate, hold on! We'll be there in a moment.'

I stopped them plunging in, but as they pulled me out, I couldn't persuade them that I didn't need artificial respiration. They soon had me face down in the mud, practising what they had probably learnt on a first-aid course. I wanted to stay behind to watch for Captain Ogilvie and any other swimmers, but they insisted on escorting me to a medical orderly. 'He swam the Rhine and we fished him out,' they said.

I joined a stream of shivering men, was bundled into a lorry and taken to the tents of a first-aid post. Here an orderly, detailed to give treatment for shock, tried to inject me with morphine. I literally fought him off. Finally he went for a doctor, who believed me when I said there was nothing wrong with me except that I was cold and wet.

My uniform was thrown on to a pile of soaked and bloody clothes, and I was wrapped in a blanket, sat on a chair with a little oil stove underneath it and given a cigarette and a cup of very sweet tea. It was paradise.

Later, I climbed into one of the ambulances waiting to take us to Nijmegen, where we were given a marvellous, touching reception by the cooks and staff of the Second Army, who had been hoping to join us across the Rhine. There were candles on the tables, wine glasses full of rum and the most delicious and enormous three-course supper, after which I spent twelve hours in a dreamless sleep in the barracks.

The next morning, the quartermaster fitted us out with clothes to replace our blankets and one officer, still wearing his blanket, called us to parade for a roll-call. After each name, if the man were not present, anyone who knew what had happened to him spoke up. I hoped there might be some news of Captain Ogilvie, but no one had seen him since the night before.

On Thursday, our whole division drove in one huge convoy from Nijmegen along the Brussels road to Louvain. In the late afternoon, we lined up for a combined tea, dinner and supper, and in the evening made friends with the Belgians on an extended pub crawl. By the next evening, we had been flown back to our own aerodrome in England. Our huts had been locked and left as they were twelve days ago. Four of us looked around the hut at eighteen empty beds. I thought back on what we had been through then, for the first time, and fell asleep that night happy to be alive.

A few glider pilots and soldiers from the 21st Independent Parachute Company behind Stationsweg in Oosterbeek, 22 September 1944. Louis is at the rear on the left, partly hidden. (Photo Mr A.L.A. Kremer-Kingma, Ans Kremer collection)

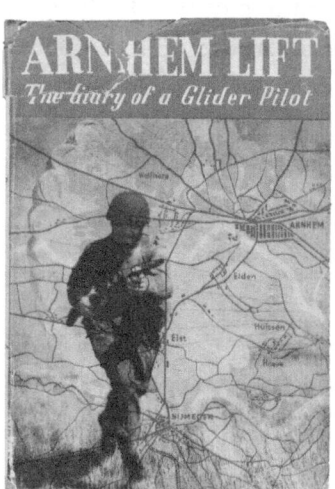

Covers of past editions of *Arnhem Lift*. Left: the 1945 first edition, by Pilot Press Ltd, that didn't use Louis' name. Right: Pen & Sword, 1993 edition.

13

The Military Medal

Late the next morning, I was called into the office of Wing Commander Lillywhite, the commanding officer of the airfield, a 6ft giant with an open, kindly face. I saluted smartly and he shook my hand. 'So glad to see you back, sergeant,' he said, smiling. Then his face became serious. 'Something awkward has happened which I must discuss with you. Of all the pilots on this airfield, they have chosen you as the only officer to be decorated by His Majesty.'

I said nothing. I didn't understand, couldn't make out whether he was serious or joking.

After a pause, during which he seemed to wait for a comment from me, he continued, 'Look at you, the way you salute, the way you stand to attention in front of your commanding officer – and what sort of uniform is this? Half-army, quarter-RAF and another quarter-gypsy.'

I managed to get out, 'I was not prepared, sir. I've only just got back.'

He went on, 'It's no matter now. The fact is, you have been awarded the Military Medal for Bravery in the Field, and it is my responsibility to get you ready for the ceremony at Buckingham Palace. It will take place in about four weeks' time. I have discussed this problem with your Army Air Corps officer and we have decided the best thing would be to send you for a course of intensive smartening up with the Guards Brigade at Wellington Barracks in London. How long do you think you need to be prepared for this occasion?'

I was still confused. I had never heard of the Military Medal (MM) and could not imagine why I was getting one. All that really sank in was 'Wellington Barracks in London', and without hesitation I replied, 'As long as possible, sir.'

'All right, that's settled. Collect your travelling vouchers tomorrow, and good luck!' Looking me up and down and smiling incredulously, he added, 'But I beg you, don't let the squadron down.'

I was told later, by Colonel Murray, my Army Air Corps commanding officer, that Captain Ogilvie had sent a radio message from the HQ at Oosterbeek to propose my award. At no time during the battle had I suspected he thought so much of me; all I felt was that he trusted me. I was deeply upset to learn that he had drowned in the Rhine. He was a good person, warm-hearted and with a great sense of humour, and I had hoped that by some miracle he might still be alive.

I slowly began to realise what winning the Military Medal really meant. A payment of £30 went with it, which was a lot of money to a sergeant glider pilot. And there was another, much more important benefit: the little red, white and blue ribbon I would now be allowed to wear would enable me to get away with scores of disciplinary misdemeanours – superior officers, especially peacetime officers, were greatly impressed by it. At the time, though, the idea of getting two weeks' survivors' leave in London, on top of my being posted there, was the greatest bonus of all.

I had many friends in London, the closest of them Vivian and Dido Milroy. I had known Vivian since before the war. He had been stage manager at the Barn Theatre in Shere. He was slim, fair, full of fun and enthusiasm and interested in almost everything, from collecting exotic fungi to wine making. He had played double bass in a semi-professional dance band that had appeared regularly at the Players' Theatre in Covent Garden. Now he was in the Army Radio Unit.

Dido came from a well-to-do country background and had large, laughing eyes and a sensual mouth. She was extraordinarily bright, vivacious, opinionated and witty, with many close friends involved in literature and the theatre. A fabulous cook, she did a lot of entertaining at their large Georgian town house in Primrose Hill. During my smartening, I stayed in their small spare room.

Like everybody with friends or relations in airborne units, Dido had followed the highly publicised progress and final catastrophe of the Arnhem operation, and was greatly relieved when I turned up alive. So, too, were all the other friends I rang on the first few days of my leave, and they showered me with invitations and affection. Wherever I went and whomever I met, I had to tell them what happened at Arnhem. I got so tired of repeating the story that I thought I would save myself a lot of trouble if I wrote it down in a pamphlet and gave it to anyone who asked.

Dido was all for it and got everything ready for me to start writing. But the telephone kept ringing, unexpected visitors kept turning up and invitations for parties and dinners poured in. I had no peace, and besides, once I had spent a day or two struggling with the typewriter, I had lost my enthusiasm and was longing to go out on the town and enjoy myself. But Dido willed otherwise; she was determined that I should make something worthwhile out of my experiences and kept me at it. She answered the telephone, said I was out, wouldn't let in any visitors and refused all invitations on my behalf. For ten days – as long as the Arnhem operations had lasted – I laboriously tapped out details of my battle experience, and she corrected and edited what I wrote.

I was not easily disciplined. Temptations of the flesh and the attractions of the outside world were too powerful for me to remain shut up during my precious leave, so Dido devised a rule by which I was allowed to indulge the necessities of my life, love and food, only when three pages had been completed. I tried to fool her by enlarging the white margins of the typed sheets and increasing the spacing of the lines, but she always noticed and was adamant.

When my leave was over, I had to report to the Guards Brigade at Wellington Barracks. The first thing they did was issue me with a smart new uniform and assign a corporal to give me tuition in military bearing, marching, saluting and keeping my boots, belt, brass and buckles up to the highest gloss. Otherwise, they didn't really know what to do with me. I don't think they dared risk assigning me to guard duty at Buckingham Palace or St James's Palace, so they left me to my own devices. I soon got to know who mattered in the regimental office and, with a little bribery, got myself excused from all duties except the weekly pay parade.

I had a lovely time, with the Medawars in Oxford and the Milroys in London. After almost a month of strenuous weekly pay parades, I was given details of the investiture. I was to report at the Wellington Barracks parade ground at ten in the morning, in the company of my parents, wife and children, if I had any. At precisely 10.18, I drove through the gates in Peter Medawar's rickety old Standard 8 in the company of Charlotte, an attractive girlfriend from the American Office of Strategic Services, Dido and my godchild, Caroline. We were in the middle of a procession of smart chauffeured and beflagged limousines.

After I was told where to park, we joined the other guests assembled in little groups. Ours was the smallest; others consisted of senior officers in dress uniform with wives and children. Soon, a parade formed up in front of us and started marching towards Buckingham Palace. All those watching stood to attention or saluted. I followed suit, until I noticed with astonishment that those parading in front of us were army air force officers, NCOs and privates all mixed up. Only then did it dawn on me that this must be the group who were to be decorated at the palace by King George VI. I quickly broke into their ranks – there was some shuffling and quiet cursing as the parade adjusted to absorb me.

Just as things quietened down, Caroline came running after me. 'Büdi, I must pee!' she said. I broke ranks, was cursed again and took Caroline to the men's toilets, which she found a very interesting new experience.

By the time we got back, everyone had disappeared into the palace. We hurried after them and told the guardsman on duty at the gate that we were part of the investiture ceremony. He pointed towards the many doors of the palace, saying casually, 'You will find them somewhere in there.' We crossed the empty courtyard in front of the palace and entered the wide door indicated by the guardsman.

There was no one in sight in the large, bare entrance hall, and not a sound to be heard, so I went cautiously on, my little group silently tiptoeing after me through the corridors, empty offices and storerooms. It was not at all as I'd imagined a palace would look.

At last, we heard faint voices and made our way to a large, opulently carpeted room where a crowd of officers and other ranks were mingled together, chatting noisily. The major-domo, a tall, elderly gentleman most elegantly turned out in a frock coat, came up to me and asked, 'Are you being awarded the Victoria Cross by His Majesty?'

'Certainly not, sir, only the MM,' I replied.

'But how is His Majesty to know this?' he asked.

I shrugged. If he didn't know, I certainly didn't. 'Look at all these men,' he continued. 'They have their ribbons pinned to their chest. Where is yours?'

Before I could reply, he summoned an adjutant and told him, 'See that this man's family is taken to the visitors' gallery as soon as possible.' Turning back to me impatiently, he asked, 'What is your name, sergeant?'

'Haig, sir!'

'Damn it, the beginning of the alphabet. Hurry up and put your ribbon on; it will be your turn soon.'

'I haven't got one – I thought I got it from His Majesty,' I said.

'Didn't anyone tell you anything?' he asked, and called into the milling, chattering group, 'Anyone with a spare MM ribbon?'

Someone came forward with a 30cm ribbon. The major-domo rolled it up and pinned it to my chest. As he stepped back, looking me over with disapproval, a voice barked out, 'Sergeant Haig!' It was my turn, and I was gently pushed through a large, arched double door into the ballroom, where the investiture was taking place.

As smartly as I could, I marched the long distance across the parquet floor, my heavy army boots sounding through the hall. I halted, stamped my feet noisily and saluted in front of His Majesty, just as the Guards corporal had taught me.

The King was not alone. The Queen and his two young daughters, Elizabeth and Margaret, were with him. An adjutant handed him the large silver medal on a velvet cushion. He pinned it underneath the rolled-up ribbon, shook me by the hand and congratulated me. Then I shook hands with the Queen and the two young princesses. 'And where do you come from?' the King asked.

Before I could think it over, and regretting the words the moment they slipped out, I answered, 'Potsdam, sir!'

The King stammered out something I couldn't understand. Then, after an awkward pause, during which his jaw moved but he didn't make a sound, he continued, 'Arnhem – a terrible tragedy – so many men – but I am sure you are keen to get out there again soon.'

I was mesmerised by the poor man's struggle to get out his words, and before I knew what I was saying, I answered, 'No, sir, I've only just got safely back.'

From the look in his eyes, I realised the interview was over, and I stamped my feet again, saluted and marched out of the hall. Later, my friends told me how the Queen had stared at me, unable to understand how a German came to be in her drawing room.

A few days later, Dido produced a neatly bound little manuscript, which I handed to my commanding officer, Colonel Murray. Army regulations stated that anything written for publication had to be approved by him. Two days later, he called me into his office, where the manuscript lay on his desk. He angrily confronted me, 'You should be ashamed of yourself. No Britisher would ever have embarrassed his comrades by writing stuff like this. Your book lets the whole regiment down. I won't even pass it to the War Office. I forbid you to contact either press or publisher. Sergeant Haig, I am very disappointed in you – you of all people, a holder of the MM.'

That was that, as far as I was concerned – but not for Dido. When she heard about the colonel's outburst, she sent her copy of the manuscript to one of the chief censors at the War Office, who passed it with the comment, 'The manuscript is excellent and suitable for publication. The time is right when we can afford to show the seamy side of the war from the viewpoint of an ordinary man, rather than all this professional stuff from reporters.' His only proviso was that we must delete anything that might give away the fact that the author was German. So the book was published anonymously and called *Arnhem Lift: The Diary of a Glider Pilot*.

The Military Medal

Finding a publisher was more difficult than getting War Office permission. Mumpitz, my old friend, had become a literary agent and telephoned masses of publishers, but none had any paper left from their years of rationing. Only the brand new Pilot Press still had an allocation of unused paper, and as the book had only a hundred pages, they decided to publish it. It went into many editions, was translated into nine languages, and became the basis of a film called *Theirs is the Glory: Men of Arnhem*.

But before any of this happened, and soon after my interview with the colonel, I was transferred to a glider pilot squadron in India. I felt sure the colonel had been responsible for this. He'd wanted to get rid of me, and now he had succeeded. I was very upset. I wanted to be there when the European war was won and Nazi Germany defeated. I had no wish to fight in or for India, or even to be associated with the imperial side of British history. To me, at the time, there was nothing attractive about the Far East.

Medals.

14

India

India was about the only country in the world I had no desire to visit. I thought I knew much about it and I did not like what I knew. I believed Britain had exploited the land for almost 200 years, but that the people had nevertheless had the guts to revolt again and again in an effort to throw off the foreign yoke and build a free and united nation. And as the British had ruled India for so long, surely its present deplorable state was entirely their responsibility? If 95 per cent of the population were illiterate, why hadn't the authorities taught them to read and write?

This was what I thought when I got out of the luxurious Sunderland flying boat at Karachi, and it was an outlook pretty typical of the average-thinking soldier. But the more I saw and experienced during my fourteen months in India, the more I changed or reversed my opinions – and the harder it became to draw any sort of conclusion.

There can be no greater contrast between travelling in a Sunderland and sitting in an Indian train. We had taken barely three days to travel from Poole harbour to Karachi in a VIP atmosphere, lounging in easy chairs, strolling to the larder or icebox for refreshments, playing cards and chess and idly watching the pattern of the world unfold beneath us.

Karachi's railway station reminded me of Paddington Station in London on a busy bank holiday. How many of these Indians ever intended to, or actually did, travel is difficult to say: they were just there. They squatted and lay on every inch of a platform already covered with refuse, spit and excrement; among them bald, sore-covered pye-dogs sniffed and scavenged. The atmosphere was hot, fetid and airless. A loud buzzing noise rose from the crowd, punctuated by the wailing of babies, the sharp yelping of the dogs as they were kicked, the shouting of coolies and the excited voices of people quarrelling with each other. Dotted about the station were mysterious bundles of white sacking; luggage, one thought, until one noticed a tiny pair of eye holes about 2ft off the ground, belonging to women who had been dumped in the blistering heat and forgotten.

Whole families remained for days together, waiting for a train that might be able to take them somewhere. They spread a rug in the dust and settled down to cook, eat and sleep. Sometimes they came because there was a roof to shelter them from the sun or rain; many came to beg or to run one of the many portable food stalls. Gangs of wild children roamed the platforms, either naked or dressed in filthy rags, their long, lice-infested hair hanging over their shoulders. If any food was spilt on the ground, they would reach it faster than the pye-dogs.

Most of the professional beggars were women who invariably carried a baby in their arms. They followed you wherever you went – '*Baksheesh, sahib?*' – plucking at you with greasy, claw-like hands until you were nauseated. In the front rank of the beggars were the crippled, the blind and the syphilitic, being led, wheeled or carried.

Once, when I looked out of the train, I saw something being carried level with the window in a small basket. It was a tiny, withered body, the size of a 2-year-old child, on top of which was the frail, oversized head of an old woman. Through the open window of the compartment, the basket was pushed towards me and in a pathetic, cracked voice, the head monotonously repeated, '*Baksheesh, sahib? Baksheesh, sahib?*'

The first time I saw this sort of thing, it left me stunned. To start with, I was overcome with pity, which changed to irritation, then into something very near hatred. I hardly noticed the change at first, except that I started turning away to avoid seeing the beggars. Then I found myself actually pushing those dirty bodies away when they came near me or touched my clothes. I was horrified to find myself shouting and losing my temper. I was ashamed of my irritability; I knew perfectly well it was wrong.

The train was more than two hours late when it finally pulled into the station. The Indians became quite frenzied, stampeding towards it, elbowing each other and kicking aside anything that happened to be in their way. Their carriages filled almost at once, but more and more people pushed and shoved their way in, until there were layers and layers of humanity piled up to the roof. Some passengers climbed in through the windows and were levered into place by friends and relations who may have come to the station especially for this purpose. When the train eventually started, after a great deal of whistling and flag waving, there were arms, feet and whole bodies still hanging out of the windows and doors.

We had our own reserved compartments – the height of luxury compared with what the wretched Indians had. But to those of us who were going to have to spend five days and nights in these stinking, wooden-benched carriages, comparisons weren't much comfort. Besides the general heat and overcrowding, the place was alive with mosquitoes and we were soon overrun by formations of dark red cockroaches that appeared from underneath the benches and the latrine. Ants, too, found their way into our luggage and insects buzzed around us day and night. We had no place to wash and the toilet, which stank, was just a hole in the floor. We felt even grimier than after days of fighting. We were beyond feeling dirty – we felt unclean.

As time went by, we got more and more tired and depressed. During the day, the heat made sleep impossible, and at night we were kept awake by the mosquitoes and by the crowds of Indians who would try to sell us tea or fruit every time the train stopped at a station or siding. If they got no response, they went on clamouring and banging at the shutters, trying to force their way into the compartment. Any man not already roused by them would be woken by our cursing.

Three times a day we stopped for food. While the officers went to the station refreshment room, rations and tea were distributed among the lower ranks. Children and beggars lined the train on both sides and fought over every scrap we threw out of the window. We never saw anyone refuse anything, even meat.

The first-class sleeping cars in which the officers and wealthier Indians travelled were very comfortable, fitted with baths, fans and sometimes air conditioning. As the journey went on, the officers from our draft devised a scheme to better our lot. They unpacked spare tunics, and whenever a passenger got out, a newly promoted officer promptly got in. It was a great moment when one's turn came to take a temporary commission and have a wash and a refreshing nap under the fan, while for those who were still waiting, this at least made our own carriages less crowded.

Apart from the seething activity on the railway stations, there was very little to see on the journey. We travelled through almost 2000 miles of yellow-brown, monotonous wasteland, as bald as the back of a mangy dog. The cattle were bony and undernourished. The primitive mud huts and hovels had no amenities, and naked children and ragged, lean men and women were always moving in and out. That journey, the first of many, opened my eyes to the size of India and to the desperate poverty of most of its people.

After almost a week on the train, we arrived at our camp – a collection of huts set down in the barren landscape. Once we had settled in, each day was like every other. It was not a strenuous life; there was no guardroom, no fence around the camp, and within reason we were free to do what we liked. Yet there was always an atmosphere of confinement. The time after tea was most trying. At about five o'clock, when the light was fading, we got restless and longed for recreation, for a change of surroundings, for female company, for family and friends. We felt the urge to dress up, go out, lead our own lives, do something different from the daily routine. We couldn't help remembering how it had been after tea at the camp in England when we got up from the table still chewing, raced to the wash house and scrambled into our best battledress, already visualising a waiting girlfriend, a dance, a film or a telephone call home. Once past the guardroom, we had been free, but here there was not even a guardroom to pass.

Four of us decided to visit the village near our camp. It was supposed to be out of bounds after dark and we had been told it would be dangerous for us to go there, but in our present state of boredom, the element of danger was just what we needed. There was another reason for our excitement: we had all noticed a few huts in the village from which girls in brightly coloured silks had smiled at us. They were quite different from the tired-looking village women who did all the heavy manual work and covered their faces whenever you looked at them. There was a gentle freshness about these girls that had made us return their friendly greetings and secretly hope for more. What we were contemplating was against all the rules and regulations, but that didn't worry us.

After dark, we crossed the rough fields surrounding the village, without touching the road leading into our camp. It was pitch black but by no means quiet: the jackals and hyenas were just beginning their nightly concert. As the sounds of our camp grew fainter and the noises of the village louder – the slow whining of native music, the dull rhythm of drums, the calling of voices – so, too, the smell of human excrement that surrounded every Indian village grew stronger. We slowed down; it all seemed so foreign and unreal. But to turn round and go back to our dreary camp would have been cowardly. So on we went.

Once in the village, we strolled around examining the bazaars until, at the entrance to a whitewashed hut, we saw two of the dark beauties we had been looking for. We hesitated. It felt as if we were standing on a high board, not knowing the depth of the water below. But having got this far, there was no going back. As the girls smiled sweetly, we went towards them. They were barefoot and dressed in sheer material that hung loosely around their well-shaped bodies. When they smiled, their eyes and teeth shone in the lamplight. Gently, they drew us into the hut. We looked at each other, pleased and rather surprised; it had all been so much easier than we had expected.

We talked and laughed with the two girls; none of us understood a word of the other's language but we knew we were all thinking about the same thing. Then one of them called out and a boy of about 12 came in. He spoke English. 'Jig-a-jig, *sahib*?' he said, and holding up one hand he added, '5 rupees.'

We said, 'Jig-a-jig 5 rupees all four of us?'

He laughed and told his sisters, and they laughed, too. Then he said very seriously, 'Jig-a-jig 5 rupees each, *sahib*.'

We said, '3 rupees each, *sahib*, and you're a cheeky boy.' We paid him 3 rupees each, and he produced two more smiling beauties who knew what it was all about. Each girl took one of us by the hand and led us away.

Hand in hand, my girl and I crossed a dark courtyard, sinking into the mud in several places. We came to a very small room, barely a cubicle with whitewashed walls. There were no windows and no furniture, just a straw mat on the floor. The flame from an old tin filled with oil gave a faint, flickering light. Looking at the strange young creature in this dim, empty little room, it suddenly came over me just how far away from England I was. Then I forgot about England.

Back in camp, with a nightcap in our hands, we sat on our beds comparing notes. We were absurdly proud of and pleased with ourselves. I had the same feeling of exhilaration I'd had after the patrol behind the German lines at Arnhem.

Some weeks later, we had our first encounter with the British natives of India, those legendary *pukka sahibs* and their wives, the even more *pukka memsahibs*. A hundred of us were driven in trucks over dusty, bumpy roads to attend a dance and supper at the governor's residence. The 100-mile ride was a nightmare, leaving us covered in dust and aching all over, but as soon as we had passed the smartly saluting guard at the gate and had our first glimpse of Government House, we felt it had been worthwhile. Coming from our primitive huts at the station, this world of green lawns, flowerbeds and palatial buildings looked like a stage set. Fountains played and immense trees shaded the gravel paths that led through the peaceful park; everything was in complete contrast to the parched, yellow countryside through which we had just come.

The trucks slowly rounded the carriage sweep in front of the residence and drew up under a large portico, through which we went up wide steps into the entrance hall. The rooms were colossal, painted white and very simply furnished. The floors were covered with bright Indian carpets and the barefoot servants were dressed in smart red liveries.

Slightly dazed, we entered the cloakrooms and looked at ourselves in the polished mirrors. Our faces were black, and after a few minutes, so, too, were the towels. When we emerged suitably scrubbed, we were escorted to an enormous ballroom where the governor and his lady received us, he resplendent in diplomatic regalia, she in long evening dress. Around the dance floor sat the girls, assembled into several groups, apparently determined by the shades of their skin. The smallest group was the whitest; there were only five of them and they stood out like stars. None of us had seen a white woman since we had arrived at the camp. Some of the men just stood and gaped but about twenty of us, myself included, made a beeline for the five. Within a few minutes, competition had forced me to show off my medal and say I had written a book. This got me shamefully satisfactory results. I was now one of a group who were considered sufficiently 'all right' to be led away and shown off to the governor's party.

When we were introduced to members of the smart set, we found we each had a label. Sergeant Haig, MM, famous author (that was OK by me); Flight Lieutenant Lestrange, journalist (Tony was 19 and had once had something published); Warrant Officer Slade, Battle of Britain Spitfire pilot. We knew how to play the game now and enjoyed it, in a rather awful way, but the *memsahibs* took their snobbery deadly seriously. If I asked about some particularly attractive girl, I would be told, 'Oh, she's country', meaning mixed blood, or, 'She's married to an Indian', or, 'Her father works on the railway.'

Before the evening was out, I had been told about the private lives of all the girls who 'mattered'. It seemed they had nothing else to do but sleep with each other's husbands or with an odd lover now and again. That was what they told me about each other, but I suspected their lives were a great deal tamer than they wanted me to believe. They all had at least ten servants, cars, *ayahs* and hard-working husbands. Their idea of war service was to roll bandages over a nice cup of tea out on the lawn, to dance with 'our boys' once a week or to walk through the hospitals for an hour or so and say a few words to the wounded soldiers, who were assumed to be very grateful.

Their lives were a well-ordered round of lounging under cooling fans in their beautiful homes and meeting in the clubs to talk scandal over drinks, always with a delicious feeling of superiority over the Indians and Anglo-Indians. But they were women and they were white, and we had come to the governor's beano with the firm intention of cadging invitations. So when one of my partners, Felicity, seemed kindly disposed towards me, I ventured to suggest a date. She promptly invited me to lunch.

My first visit to her large white house was more or less formal, but my hostess made it clear that from then on she would be at home – 'On, should we say, Wednesdays?' – if I cared to call. I did care, even though I discovered that to spend three hours with Felicity would involve twenty-four hours of the most excruciating travel.

The first thing I had to do was to wangle myself on to the mess committee as a buyer of fresh food. This meant I could leave the camp early on Wednesday morning, reaching the market town several hours later. I was filthy and worn out before I got out of the *gharry* (a horse-drawn cab), which returned to camp with the shopping. My work done, I wandered about in the dust and dirt, as the day grew steadily hotter, until two o'clock when I could at last ring my *memsahib*.

'Yes, you can take a *tonga* and come now. He's just leaving for the office.' That was the shameless answer I had been waiting for. Ten minutes later I was let in by one of the uniformed boys.

She was waiting for me in her air-conditioned boudoir. Outside it was 110 in the shade and the brilliant light hurt my eyes; inside the temperature was kept under 75 and the blinds were drawn. It was a small room; a large couch upholstered in pink chintz covered most of the floor space, with a small easy chair and a large cocktail cabinet, and that was all. A door led into a little bathroom and the bathroom conveniently led into the garden.

Felicity, draped on the couch, was pretending to read my book. 'Darling, you are a brave boy, aren't you?' I smiled, trying to look suitably modest. 'Do you think you'll write a book about India? If you do, will you put me in it?'

I said I thought, on the whole, that I would. She gave me a long, ice-cold gin. I never got any whisky there and later found that was the only concession she made to her husband: the whisky was his. She settled me in the chair and began to talk.

I felt so wonderful, sipping my long drink and breathing the cool, clean air, that I hardly listened, but I understood that everything she said, her catty criticisms and naive boasting, was designed to impress me. What did impress me was her beautiful body and very pale skin. Before long she was on to her favourite topic: 'The things I do for the boys.' And by gradual degrees her conversation became more and more personal. It was the kind of talk, punctuated with giggles, I had heard so often at small-town dances in England, and there was only one way to stop her.

At five o'clock, her husband rang to say he was coming home, and I had to go.

Coming out into the dry heat was like entering a furnace. I could hardly breathe and the sun blinded me. I walked down the wide road until I came to a railway crossing, where I waited for a slow goods train. I jumped on the brake wagon and got a lift back to the goods yard. Here, I changed to a locomotive, which took me another 18 miles to a cinema that was patronised by the local RAF. *Gharries* from their aerodrome came to this cinema every night and after the show I got a lift back to the RAF station. At five o'clock the next morning, I hitched back to our camp on the mail plane.

Looking back, it seems incredible to me that those three hours of pleasure with Felicity were worth twenty-four hours of extraordinary discomfort – but they were. In spite of her egotism, her vanity, her lust for admiration and attention, and the coldness that showed itself in the way she treated her husband and her servants, she did at least do something for one of 'our boys'.

In the middle of June, the heat became unbearable. The sky had been cloudless for weeks, there was no breeze, and even at night the thermometer registered more than 100 degrees. We woke in the morning, sodden with sweat, tired and parched. I don't know how long this lasted because I lost all sense of time.

One morning, I felt I could bear it no longer. I went to the commanding officer and asked for leave. He saw no objection and three days later I was on my way to Kashmir with a friend. The journey took six days, and by the time we reached Srinagar, at 2,000m, it was like coming out of anaesthetic and breathing fresh air again.

We had booked a houseboat by telegram and its owner was waiting for us with a horse-drawn carriage. The boat was delightful, built of varnished wood and fitted with an awning, under which deckchairs and tables were arranged as though for a tea party. A little carved wooden fence ran round the deck, on which pots of geraniums were blooming. There were three large bedrooms, a pantry, a dining-cum-sitting room furnished with bright Kashmiri carpets and curtains, comfortable furniture and flowers everywhere. Our staff of four servants lived in a small domestic houseboat attached to ours.

We began to feel at home, even more so after the first hot bath we'd had for months and, following the unending cycle of tinned rations, a dinner of succulent roast duck, fresh cauliflower and peas, fruit and creamy custard. After that came real deep sleep, wrapped in three blankets and fanned by cool air.

The next morning, the houseboat moved on into one of those strange, blurred landscapes that grow with their roots in the water; flooded fields and floating gardens all seemed to swim in a haze of liquid green. We found a quiet mooring place on the Nagin Lake and settled down to swim, ride and explore the other lakes and the wonderful, centuries-old Shalimar Gardens, built by the Mogul kings on their way down to the plains of India.

Ten days later, we had recovered our strength, and along with it our curiosity. We were no longer satisfied with our lazy life on the lake. We decided to go on to Gulmarg, a small hill station about 30 miles away and 2,700m above sea level. I felt as if I had been there before; the atmosphere reminded me of Switzerland, and of skiing and climbing in the high Alps, the mountain air inducing a feeling of irresponsibility. The beauty and danger of snow is hard to beat.

We rode up through tall trees, our ponies picking their way along a rough track leading to a little plateau at about 3,300m. Here, we dismounted and started climbing. The grass grew shorter and thicker, and was starred with brilliant alpine flowers. We reached the snow and began to climb up the steep icy crust, sliding and stumbling, and often losing more height than we gained. As the crust became thinner, we broke through it, sinking knee-deep into soft snow. As the air became thinner, too, we began to puff and pant. Every fresh step seemed to use our last ounce of energy, but there were still another 300m to go.

At last we stood at the top, feeling dizzy and exhausted, but with an indescribable sense of pride and satisfaction. The white sheet of a colossal frozen lake lay before us, surrounded by the myriad white peaks of the Himalayas. We felt on top of the world.

The next day, we tramped through the deep snow with sledges on our backs. Still out of breath, we tried to calculate a course down the 1,200m descent to green grass below. Then, tightly perched astride the sledges, we took the plunge, braking with our feet as we began to move faster and faster. The snow from our heels sprayed up into our faces; we tried to steer but our eyes were clogged with flying snow.

When a boulder loomed in front of us, there was nothing to do but fall off and roll and roll through the soft snow. We watched each other come down, each laughing and struggling. Finally, down on the grass, soaking wet, our faces and hands stinging from the cold, we felt wonderful.

Sitting on the terrace of our hotel later, our clothes drying in the warm sunshine, we drank tea and ate wild strawberries and cream. Then we danced all night in the warm wooden annexe of the hotel, delighted with the pleasure of holiday friendships and pretty girls in evening dress.

At four in the morning, we had a breakfast of eggs, bacon and strong coffee, before riding back to the houseboat at dawn, reaching it as the first light turned from sapphire to ruby red behind the outlines of the Himalayas.

Sitting in a waterfall,
Kashmir, 1945.

In the old Mughal
Garden, Kashmir, 1945.

15

Becoming a Journalist

From the military point of view, our first six months in India had been spent idly and we had constantly to remind ourselves why we were there. So it was a great relief when things finally began to move and we were posted on a succession of courses to different parts of the country.

One day I was asked whether I would like to become a pilot officer. As this promised new scenery and a change of routine, I agreed willingly, never expecting I would be sent to Poona, the most traditional of all the empire's officer training establishments, and the source of innumerable music hall jokes about Colonel Blimps. It was a pleasant surprise to find that life at the training academy was civilised. We had spacious sleeping and living quarters, beautiful gardens and each had a batman to look after us. We were treated like gentlemen and our training was interesting, except for the morning parades: at six each day we had to fall in and execute the most complicated marching manoeuvres on the huge parade ground. It was a rehearsal for the march-past each Sunday morning, when a high-ranking officer, usually a general, took the salute. Still half asleep, I always found a place in the middle, where I had people in front, behind and at each side, so that when we marched straight ahead, swung left and right, turned about, split ranks and joined up again, I could do it like a sleepwalker, without thinking.

One fine Sunday morning, we assembled as usual around the fringes of the parade ground. The general's rostrum was whitewashed and red carpeted. A Scottish guardsman in a kilt groomed the regimental goat to lead the parade, and pipers were tuning up their bagpipes. I had climbed a mango tree to see if I could reach some fruit when I heard my name being called by several of my fellow cadets.

'Lewis!' they shouted, 'The RSM wants you!'

I climbed down, straightened my uniform and reported to the regimental sergeant major, stamping my feet, saluting and shouting, 'Sir!'

'Sergeant Haig,' he said, 'you are going to be chief marker for the parade today.'

'Oh no, sir,' I blurted out. 'I couldn't possibly.'

'Don't talk nonsense, Haig. The US general who was in charge of the Americans at Arnhem is taking the salute. He has been told about your exploits there and your MM, and he wants to meet you after the parade.'

I entreated him to choose someone else. He dismissed my objections, telling me not to be so modest and that it was an order, before walking smartly towards the rostrum where the general was expected.

Before I had time to dwell on my desperate situation, the bugle sounded and the RSM shouted, 'Marker out!'

I marched out in front of the parade and led them forward. They followed. I thought I remembered that a right turn came next, followed by a left. After that, my mind went blank and I led the parade at random. I was fascinated by the certainty that quite soon the situation would blow up. It felt like the moment before a car crash, when one sees the disaster happening in slow motion, unable to do anything about it. I had already led the parade into two more right wheels, so that it cut off its own tail, in complete disarray. The cadets were doubled up with laughter and couldn't hold their rifles upright. The bugler sounded the retreat and the cadets ran from the parade ground while I fled into the mangroves.

The RSM reorganised the parade, bellowing his commands; the bugler sounded the fall-in and the show started off once more, this time with a new and experienced marker. I stayed hidden, wondering what my fate would be.

As it transpired, I was not introduced to the American general but, to my astonishment, neither was I arrested; the officers totally ignored me. I was certain, however, that this incident would not be forgotten and waiting for my punishment was terrible.

It was three days before the commanding officer summoned me. He got up slowly from behind his desk and said, in a quiet, almost resigned voice, 'Sergeant Haig, we have thought over and discussed your behaviour. What you did was so disgraceful that there is no precedent in the 200-year history of this college. We have not yet been able to decide on a punishment, but you can be sure that you will hear from me in due course. In the meantime, I never want to see you on the parade ground again.'

From that day on, I was excused from every parade. But apparently, no fitting punishment was found for my misdemeanour, for I never heard any more about it. I think there were two reasons for this: I had been decorated, and I was the only enemy alien in Poona, maybe the only German ever to be trained there.

I was now qualified as an officer. My papers had not yet come through, but instead of waiting for them, I decided to go to Calcutta (now Kolkata), where I knew officers' uniforms were made cheaply and well.

After I had found my quarters and been measured up by the tailor, I planned to celebrate my promotion with a drink in the Grand Hotel – but they wouldn't let me in because I wasn't wearing an officer's uniform. I tried to convince them I was newly qualified, but I had nothing to prove it. At that moment, a captain came by and asked what the matter was. He settled the argument by saying I was his guest and buying me a drink at the bar.

Noticing my wings and glider pilot epaulettes, he asked me about Arnhem and introduced himself as Ian Coster, editor of the illustrated *Phoenix Magazine* and the SEAC newspaper, both published for English-speaking troops of the South East Asia Command. He showed me the current issue of *Phoenix*, which looked very much like America's popular *Life* magazine, and was a favourite of the hundreds of thousands of troops in South East Asia. Featured prominently in it were extracts from my book, *Arnhem Lift*, just published in England by an 'anonymous author'.

At first, I was too excited to say anything, but at last I managed to point out the incredible coincidence. 'You may not believe me, but I wrote *Arnhem Lift*.'

Coster didn't let on whether or not he believed me, but he questioned me closely. After an hour or so, and a great number of beers, he got up and said, 'Let's go to the office and have you photographed.' This photograph appeared in the next issue, and a great story was made of the discovery of the author of the book about Arnhem. I also met the assistant editor of *Phoenix*, Tony Clarkson, a naval lieutenant, and Hugh Cudlip, editor of the SEAC newspaper.

The next day, we all went out for dinner in Calcutta's Chinatown, and by the end of the evening they had decided I must join *Phoenix* as a roving reporter. Next morning at breakfast, my story was related to Colonel Frank Owen, a well-known English journalist who had once been the youngest Member of Parliament. When I told him I would love to join his outfit but thought the Glider Pilot Regiment would never release me, he said, 'Leave that to me, Haig; I am going down to the HQ at Kandy and will get it approved by Lord Louis.'

Less than a week later, Tony Clarkson handed me a little red pass, certifying that 'Lewis Haig' was a correspondent for *Phoenix Magazine* and must be given every assistance in obtaining information, transport, billets and finance. It was signed, 'Louis Mountbatten, Commander in Chief, South East Asia Command'.

I went to the tailor and got him to change my pilot officer's insignia to that of war correspondent, then started immediately as a reporter on *Phoenix Magazine*. Here, although the staff wore the uniforms of all three services and ranked from private to colonel, we lived as one happy family, eating and drinking and working together as equals.

Tony Clarkson was my direct superior and gave me my assignments. My first was on the dangers of visiting brothels, warning the naive and sex-starved members of His Majesty's Forces of the consequences of such visits. Tony was an excellent journalist and had no difficulty in finding highly moral excuses to present the more sensational aspects of service life in India, illustrated with lurid pictures. Nor did we have any trouble finding volunteers to play the part of clients in a brothel, all set for the critical moment when the Military Police, accompanied by our photographer and a few journalists, would burst through the door.

The magazine also ran features on Indian music and religion, the Sikhs, the Anglo-Indians, the all-Indian kite-flying championships, twenty-four hours in the life of an Indian village and the beggars of Calcutta.

I could never have managed without Tony, being completely inexperienced. Until *Arnhem Lift* I had not even written a long letter. But however horrible my copy was when I handed it to him, he transformed it at lightning speed. His hand raced across my typed pages, making alterations, transferring words or sentences from one place to another, adding, crossing out and punctuating. I felt like a fraud when my name appeared under the printed article, but every time I handed in a story, apologising for its state, Tony would say, 'As long as your copy is as meaty as usual, don't worry – the editing I can do in my sleep.'

Through my work as a journalist, I got to know Calcutta very well. The West had cultivated and forced it into expansion, without any thought beyond the present. Several thousand white people lived and minded their own prosperous European businesses there, quite unconcerned about the native millions in the rest of the community. It was a place where great riches flourished beside great poverty. I found it impossible to get used to seeing a half-starved mother and her child camping on the hard pavement beneath the window of a rich man's house, or a family of ten crammed into one tiny room. Millions of Indians were hungry and would never have enough to eat, and the sight of the beggars, the crippled and the diseased lining the streets shocked me a hundred times a day.

The European part of Calcutta occupied less than a tenth of the city. It looked European enough until I left my car and began to walk. Chowringhee was Calcutta's Oxford Street, but there anything and everything, from parrots to pornography, tiger skins to children's toys, musical instruments to monkey nuts, was sold not in the shops but on the street. Crowds of Indians accompanied me wherever I went, offering their goods, cajoling, setting prices and reducing them every 10m. If I carried a bag or parcel, it would be wrenched out of my hand by a coolie or a child who wanted to carry it for me. If I happened to walk slowly, a sly-looking character would attach himself and offer to fix me up with a 'proud college girl, very white, very young'. Every few yards along the pavement, a boy with a tin flute played *Deep in the Heart of Texas*, hoping I would buy the instrument just to get rid of him.

There was plenty of cheap entertainment in Calcutta. The European cinemas showed all the latest films, and in the hot season we went to whatever was showing, as much as anything to enjoy the clean, refrigerated air. We might have been at the Empire in Leicester Square, except for one thing: the perpetual sound, from the cheaper seats, of Indians fighting their universal catarrh. This noise, like the twittering of birds, woke me every morning, then continued throughout the day. We got used to it. Hawking and spitting was the most common sound in Calcutta, and in the Indian cinemas, where native talkies were shown, this noise frequently drowned out the sentimental whispers of the soundtrack.

The cinemas were always packed and long queues waited patiently for hours to get in. Each film ran for years, and the audience cheered heartily, booed, hissed and enjoyed each performance. The few women we saw, daring and heavily veiled, were always accompanied by men.

Further down Chowringhee lay the residential quarter of Europeans and rich Indians. Here were larger houses, gravel drives, trees and flowerbeds, servants in white liveries and turbans. *Ayahs* pushed rosy-cheeked babies in prams, uniformed chauffeurs took care of big cars and sprawling club buildings were surrounded by tennis courts. This was probably the only quiet spot in the whole city.

Our mess was on Central Avenue, near the impressive Statesman building and only a few metres from a large refuse bin, through which every morning beggars would search for food. Its stink polluted the entire area. All day, birds, cows, pye-dogs and beggars endlessly stirred and re-stirred the rubbish; at night, scavenging rats tunnelled through it.

We soldiers and airmen-in-training had been told little or nothing about India in advance – we were left to find out for ourselves. Art, religion and culture were at first closed books to us; we knew nothing except what we saw, and what we saw was that everything Indian was inferior to what we had at home. Trains were late, uncomfortable, and slow; meat was tough and tasteless, milk invariably watered, eggs half-sized, vegetables over-ripe, unripe or musty. Beer, if you could get it, was gassy and weak, and consumer goods were of poor quality and workmanship. Everything was made as cheaply as possible, yet was almost always more expensive than British or American imports.

All this led many of the troops to feel superior and to behave in ways that would have been unimaginable in England. They didn't consider Indians to be human beings, and delighted in using the power they had, always shouting, swearing and swaggering. They drove their *gharries* at full speed through crowded streets without thought for pedestrians and animals. The number of Indians killed or injured in our village alone was appalling. But none of the drivers responsible for these crimes ever got more than a warning; after all, they had only killed a cow or pig or Indian.

The Indians could and did hate us as a race, as a matter of principle, but the hatred did not extend to individuals. While I was on the staff at *Phoenix*, I was often sent to ask favours of the *Hindustan Standard*, the *Star of India* and the *Amrita Bazar Patrika*, all of them violently anti-British publications. Having read their attacks on us, I felt embarrassed about going to their offices and fully expected to be kicked downstairs or at least have the door slammed in my face. But I experienced no animosity of any kind; on the contrary, I could not have met with more kindness and cooperation.

To many of the Indians I knew, England was the most beautiful country in the world; the England of Shakespeare and Wordsworth, Cobbett and Hardy. They knew so much more about the country than I did that I kept forgetting they had never been there. I guessed the future of Calcutta and of India would depend on those unpretentious middle-class people. My Indian friends liked discussing politics and often blamed Britain for all their country's problems. The slogan 'Quit India' provided an excuse for failure and a promise of a great future. There seemed to be a similarity between 'Quit India' and 'Death to the Jews'. The Jews had been useful scapegoats for the Nazis; now the British in India were providing a similar excuse.

As for me, I was soon to quit India myself. When *Phoenix Magazine* wanted someone to cover Burma, Malaysia, Siam, Indochina and Singapore, writing features and taking photographs, Tony Clarkson asked me to go. I protested that I had taken only a few snaps with a Kodak box camera years ago in Germany, but he said that didn't matter. 'It will be much easier with the Leica you'll be taking with you. I'll be quite satisfied if there are one or two printable pictures in every film of thirty-six.'

So off I went, first to Rangoon, where I wrote a story about General Wingate, who had led the Chindits against the rebel hill tribes; then to Penang for two weeks, where I fell head over heels in love with a pretty Malaysian girl; then on again to Singapore.

One of the advantages of my war correspondent's pass was that I was now entitled to stay in a senior officers' mess, even though I was technically still only a sergeant. But when I checked into the mess in Singapore, I was told by the sergeant-major that all the rooms were occupied. Fortunately, at that moment a brigadier general crossed the lobby and the RSM asked whether he would mind sharing for a day or two. He agreed and, while we sat in his room waiting for the extra bed, he offered me a whisky and asked about my job. When I responded, 'And what do you do, sir?' his surprising answer was, 'I am in charge of the clap.' He was, he explained, responsible for the prevention and cure of venereal disease among troops in Southeast Asia. He told me there were more men out of action from VD than from wounds, and that handing out free condoms and punishing any man caught visiting prostitutes had no effect at all. The only thing he felt would improve the health of the troops was officially supervised brothels, but that would bring an outcry from sweethearts, wives and religious leaders at home.

Suddenly, I remembered an irritation I had felt on the journey to Singapore: an itching, burning feeling every time I urinated. I hadn't worried, thinking it probably had something to do with the wear and tear of my Malaysian honeymoon. But now I had to go again and this time it felt as if I were passing crushed glass.

When I told the brigadier, he said dryly, 'Let's have a look.' He opened my fly, took out my penis, and when he squeezed it, a thick yellow fluid appeared. As he washed his hands, he mumbled, 'Just my luck. Never get away from it.' Then in a stern voice, he ordered, 'Take your trousers down – I'll give you a shot you'll never forget. But not a word to anyone!'

I was terrified. He must have noticed because as he approached with a hypodermic syringe, he said in a kindlier tone, 'You know, Haig, you're luckier than you can possibly imagine to have been billeted with me.' As he pushed the needle into my behind, he explained, 'What I am shooting into you is a new drug called penicillin. Your gonorrhoea will be gone in a day or two. Of course, it's still very hush-hush and what I have done is strictly against regulations. This new drug will cure most infections very fast, but we have only a limited supply – it ought to be reserved for serious war injuries.' I had heard Peter Medawar talking about this wonder drug, but I couldn't possibly have imagined it would get me out of a nasty situation in Singapore.

I was cured in a couple of days, just in time to get off to Bangkok to do a feature on the Thai capital. I arrived to find myself part of the advance party moving in on the heels of the retreating Japanese. No one knew whether or not there would be pockets of resistance, and at that time only a small contingent of British troops had arrived.

At the bar in my hotel, I met a young, good-looking captain in Intelligence and we decided to explore the nightlife of Bangkok. He had got hold of a Japanese staff car and a driver, and we found several open-air dance pavilions, filled with many more women than men. Some of the girls came to our table and asked us to dance with them. They talked mostly in Thai, with a few words of English and a lot of laughter. After India, this was refreshing; there, if I ever got near a woman, she would either be too shy to talk or just giggle and hide her face behind her sari.

We arranged to meet two of them the next evening and spent several happy hours dancing and drinking cocktails. When we indicated that we were hungry, the girls answered simultaneously, 'Chinese hotel?' and took us to their personal rickshaw.

The Chinese hotel did not have the usual desk, only a large board with numbers and keys hanging on hooks. The girls chose their keys and each took one of us to a clean, whitewashed room dominated by a four-poster bed enclosed with a white mosquito net.

Lida sat me down, took off my shoes and socks, and left the room. Soon she returned with warm water to wash my feet, then brought me a bowl of delicious shark fin soup. While I ate, she knelt by my side, fanning me. Again and again, she left the room to bring new delicious dishes of crab, shrimps, duck and pork. When I could eat no more, she crawled under the mosquito net to make sure there were no insects inside, fetched a bowl of water and fresh towels, and slowly undressed. I followed her into bed. She continued to fan me while we had sex, passing the fan smoothly from one hand to the other as we changed positions. Whenever I woke during the night, she was fanning me.

One night, we couldn't find a vacant room in any of the Chinese hotels, so Lida took me to her family home. It was built on tall bamboo stilts over the river and the living quarters were reached by a rickety bamboo ladder. The rooms were all on the same level, divided by thick plaited reed hangings, with either reed curtains or strings of wooden beads for doors. Her parents welcomed me most politely and her four little brothers and sisters came in to stare at me as Lida ceremoniously served tea, sitting on the floor. I finished my tea as quickly as I decently could, finding it quite a strain to just sit there, smiling. Then Lida took my hand and, after a lot of bowing and more smiling to her parents, she led me to her room.

At first, I was embarrassed at having sex almost in public. Every sound could be heard and the whole house swayed on its stilts. But I got used to the closeness of Lida's family and was impressed by the way they treated my affair with their daughter as a natural happy event.

I stayed in the city for more than two weeks, writing a feature Tony headed 'Bangkok is the Place for Me'. It was one of the most entertaining assignments I ever had. With a car and a Japanese chauffeur at our disposal, we visited outlying temples and villages, taking the girls with us. Lida did a lot of talking to the chauffeur, but I was having too much fun to worry about it. Only later, shortly before I left Bangkok, did my friend in the Intelligence Service, over a few drinks, tell me a secret I could see he was bursting to confide: until the British moved in, my sweet little Lida had been the mistress of the Japanese commanding officer.

For my last assignment for *Phoenix*, I was sent to Indochina. The French had reoccupied a large part of the country around the old capital of Saigon and had their sights set on the north, where Hanoi was held by guerrillas who had resisted the Japanese invasion. Now that the Japanese had been defeated, they were fighting for independence, led by their Moscow-trained leader, Ho Chi Minh. The British and Americans had some sympathy for the northern insurgents, but feared the Russian connection and felt a certain loyalty to their French allies. It was a situation that could easily develop into a very nasty confrontation.

Once again, an incredible stroke of luck came my way. In Hanoi, I ran into a young Frenchman, Robert Ostier, who had been an apprentice in the Hagen Bank in Berlin in the early 1930s. My father had brought him home to Potsdam and he had become a good friend of the family. He went out of his way to help me and managed to arrange an introduction to one of Ho Chi Minh's aides. This aide turned out to be a German communist, one of the few who had survived the Stalinist purges of the 1930s. The fact that we both spoke German was a tremendous help and through him – after a few days of briefing and inspecting army units – I was at last introduced to Ho Chi Minh. He was small, wiry and agile, with an expressive face and a friendly smile. Through his German aide's translation, he assured me he would appreciate the friendship of the British and Americans, and even, he added, laughing, of the 'Imperialist French'. Then he shook my hand.

Back in Saigon, I wrote my article advocating a more flexible attitude towards North Vietnam and warning that the imposition of French rule, however liberal, would drive Ho Chi Minh's party to seek help from the Russians. Tony was happy with the article and printed it, but it was withdrawn at the last moment for fear of offending the French.

16

Returning to Germany

While I was rushing around Southeast Asia, filing my stories, I often forgot that the war was over. As far as I was concerned, nothing much had changed: I was still in uniform, almost all the people I dealt with were in uniform, and I had the same difficulties in getting transport to reach my assignments.

It was only when I was called back to Calcutta that I realised my career in the forces would soon be over, and I would have to start again at civilian life. For the time being, *Phoenix Magazine* continued, but it was clear that it would fold in a matter of weeks.

I began to consider the future. I knew returning to my career at Pressed Steel would never be satisfactory, partly because I had no formal engineering training and partly because I would always be longing for the adventure I had enjoyed as a soldier and journalist.

When I began to relax in Calcutta and read the papers, I saw what an opportunity there would be for a journalist who spoke German and had lived in pre-war Germany. For years, I had hardly thought of my youth there – it was as if it had happened in a dream or to another person. Now that I had time to think about what was happening in Germany, I felt again the pain and frustration I had gone through under the Nazis, and wondered how things were going for my friends and relations. Soon, all I wanted was to get back there.

I could not bear the prospect of being stuck for weeks on a crowded troopship, so I managed to wangle a flight home three weeks later. But just before the obligatory medical inspection, I developed jaundice. My urine looked like Guinness and I knew the Army Medical Corps doctor would spot it and send me to some dreary army hospital in India. So, at the inspection, when a group of us were given glass tumblers and sent behind a screen to provide specimens, I held mine out to a neighbour and asked if he could spare half a pint. He willingly obliged; I passed the medical and was free to return to Europe.

My flight home was not as comfortable as my arrival had been; this time I travelled not in the saloon of a Sunderland flying boat but in the cold, deafening bomb bay of a Liberator.

In England, there was no immediate need to earn money because *Arnhem Lift* had become a bestseller. The royalties had accumulated, and my old friend Mumpitz, who had been acting as my agent, had kept the money for me. I had a small fortune ready to start my first bank account.

Nevertheless, I soon found myself working again. During my time in India, I had kept notes and sent long, descriptive letters to Dido Milroy in London, partly to practise writing but also with the vague idea of putting together another book. Now I was staying with the Milroys again, and Dido made various suggestions about how it might be written. With Vivian's help, we began the task of sorting and arranging the material.

At the same time, I started planning my return to Germany. I thought the best and cheapest way would be if I could get an assignment from a newspaper or magazine. I was lucky: the *Sunday Express* had reprinted a large part of *Arnhem Lift* when it had first come out, and when I got an introduction to Dick Plummer, the paper's managing director, he said I could go to Berlin immediately to write articles for them.

Dido was not pleased. She reminded me how hard it had been to pin me down to work on *Arnhem Lift* and pointed out that I would find it much more difficult to get to work on the book if I went to Germany first. She insisted that we complete the project before I left, and so we did.

Four months later, when I arrived in Berlin, the drive from Tempelhof Airport to the hotel stunned me. Although I had seen destruction often enough on the newsreels, this was overwhelming. Rubble covered the pavement and made the roads nearly impassable. Crawling over the enormous heaps of masonry were my fellow countrymen, scratching in the rubble with their bare hands, collecting bricks for the rebuild that had already started. I wondered how it was possible for people who had thought of themselves as a master race to survive in such conditions.

The Hotel Am Zoo was one of the few places still standing in the city. It had been repaired and redecorated, and was where all the British correspondents were staying. It was situated on the Kurfürstendamm, once Berlin's Knightsbridge, close to a large square, around which were the unrecognisable ruins of famous cafes, luxury stores and cinemas.

From the centre of this square, the mutilated spire of the Gedächtniskirche jutted into the sky. This was the part of Berlin I knew best, where many of my family and friends had once lived. Now the whole district was a field of rubble. I made my way through the desolation to Zehlendorf, where some houses still stood. Here my brother KV lived in a large house with his wife Yvonne and their two children. He had been sent to Berlin by the American Office of Strategic Services to prepare the currency reform that would later wipe out rocketing inflation and curb the black economy.

While the entire population of Berlin was cold and hungry, in KV's house the central heating was blazing and food was always available. KV had a car and a chauffeur, a nanny for the children, a housemaid and a cook. In his uniform, he looked quite different from the gangly schoolboy I had last seen in 1935. That evening, they were giving a party for some of KV's colleagues. I was introduced to an American general, a British politician and a Russian officer, who clicked his heels and grunted a curt acknowledgement, before relapsing into silence when I tried to talk to him. Russian officers were always the centre of attention at social gatherings, KV told me later, but hardly ever spoke. The rest of the guests had no inhibitions about expressing their opinions, and I got some fascinating insights into how the Allied authorities were coping with the administrative vacuum left by the collapse of the Nazi state.

KV had to fly several times a week to the US headquarters in Frankfurt during the Berlin Airlift, on a return flight, the plane crashed and he was killed. His wife was not only left with two very young children, but she was about to give birth to a son. I was able to be with her during most of this difficult time.

The next day I went in search of two of my oldest friends: Woelfchen Kiepenheuer, who had first come to our house as a schoolboy of 10, and his wife Lore, whom I had last seen when working at the Barn Theatre in Shere. They had been bombed out eight times in the past two years and were now living outside Berlin in a shack near a lake. They had a bare minimum of battered, makeshift furniture, and had to scrounge and barter for food. At first, Woelfchen was uneasy with me. After a while he told me what was worrying him: he felt that, as a German, he had no right to be my friend after what had happened to me and my family at Germany's hands. It took me some time to reassure him that nothing had changed between us.

What I did want to know, though, was how someone so obviously Aryan had managed to avoid becoming involved with the Nazis. I remembered that, when it was time for him to join the Hitler Youth, his mother had alleged he had asthma and this had saved him. He used to practise having attacks at our house, so convincingly that we would fall about laughing. Woelfchen never finished university because his courses in literature were replaced by Nazi versions of culture. The same thing happened when he started to learn documentary film-making. He then turned to photography and freelance journalism.

Woelfchen had been under constant pressure from the district Nazi leader to join the party, but somehow he always came up with a convincing reason why he couldn't join immediately. In due course, he was drafted into the army but avoided becoming an officer by managing to convince the authorities that he was far too nervous a character ever to be in command of men, developing a stutter whenever he was supposed to give an order. He was transferred to the Army Film Unit in Berlin and spent the rest of the war making a movie about gas warfare. He and a few like-minded friends managed to prolong its production to the extent that it was never finished.

Our talk was interrupted by a knock at the door. When Woelfchen opened it, I saw a woman – shabbily dressed, tired and haggard – who I thought I ought to recognise. When she saw me and cried, 'Büdi!' I remembered when I had last seen her, as a glamorous, confident figure climbing out of her pink Mercedes in the courtyard of our Potsdam house. It was Baroness Mausi von Osterode, a great friend of my parents. Now she, like Lore and Woelfchen, was reduced to wandering around the devastated country hunting for food. I said I was sure I could get hold of some supplies for all of them, and promised to call on Mausi the next day.

When I did so, her flat surprised me: it was almost completely undamaged by the bombing and was stuffed with beautiful antique furniture, rugs and pictures. Mausi had been a successful interior decorator when we first knew her, and even now, in the ruins of Berlin, she had managed to surround herself with something like luxury. Before the war, she would usually arrive at our house with some famous actor or celebrity in tow. Most of her friends were wealthy and influential, but she always professed strong socialist opinions. Even when I was in the concentration camp, she went on visiting my parents, leaving her very conspicuous car parked boldly outside the house. When her boyfriend joined the Nazi Party, she left him and successfully concentrated her energy and talents on her career.

By 1933, her reputation had brought her to Goering's notice, and he asked her to decorate his grandiose home, Karinhall. They developed a working relationship, once she had satisfied him that she possessed great courage, tested by a confrontation with his pet lion. She worked for him because she loved the luxury the fees could earn her and because she loved to be among the rich and powerful, but she still showed contempt for Nazi theories. She told me most of the Nazi leaders were themselves cynical about the mythology of blood, race, soil and destiny, but as it worked to please the masses, they went along with it. Mausi said she was safe as long as Germany seemed to be winning the war, but by 1943 the British air raids were beginning to shatter Nazi confidence and anti-Nazi comments became dangerous.

One morning, the Gestapo arrested her in her office and drove her to the Alexanderplatz prison. There she was accused of being a notorious reactionary who had helped Jews and communists to escape, and had spread subversive opinions about the Nazi state. She had to sign a paper declaring that she was willing to retire to 'an old people's home', a joke name for Theresienstadt concentration camp. On the journey there, Mausi and the other prisoners were pushed, kicked and screamed at. When they arrived, they were herded into huts without beds, light or water. Mausi, pressed into a corner, counted more than a hundred figures stumbling in before the door was slammed shut and locked.

She was easily the youngest and strongest in the hut, and her courage held, even in the face of such horror. When the Nazis finally fled, as Russian tanks rumbled up to the camp, Mausi made her way back to Berlin. Here, she began to work with a friend for an association formed to help the victims of fascism. The misery of the applicants was appalling and the number was endless. At night, Mausi returned to her flat and to the 'comforts' she said she couldn't do without. 'Only those who care about beauty and have lived in a concentration camp will understand me,' she said. Later, as conditions improved, she began to pick up the threads of her artistic career and dealt again in furniture, antiques and jewellery for British and American officers. Before long, she was, both socially and financially, a successful interior designer.

What of other Germans, though? As I wrote stories for the *Sunday Express* about the black market in cigarettes or the prostitution of respectable middle-class women, it struck me that it was almost impossible to find anyone who admitted to having been a Nazi – or who felt personally responsible for contributing to the ruin of their country. I longed to find out why. I thought the answers might even make another book.

I decided to question ordinary people I had known before I left Germany, then cross-check their stories with mutual acquaintances, official records and further interviews. I had already learnt one thing from Woelfchen: that it had been possible not to collaborate with the Nazis. It was not true, as I was told many times, that it was necessary to join the party in order to survive. A non-Nazi life had been possible – a life without glory, promotion or social approval, but with honour.

I interviewed six people: a fashionable doctor, a businessman, a regular officer, a simple seaman who had become a Nazi storm trooper, the daughter of our cook in Potsdam who had joined the Association of German Girls and a schoolboy who had joined the male equivalent, the Hitler Youth. Unlike the last three, the doctor, the businessman and the officer consistently thought of themselves as anti-Nazi; they had never belonged to the party, never believed in its dogmas or worn its insignia. However, it turned out that all three had still contributed to the success of the party, and to the prosecution of the war. The doctor had become rich treating high-ranking Nazis; the businessman had greatly enlarged his empire, and his bank balance, by putting his factories and his managerial skills at the disposal of the regime; and the officer had remained blindly loyal to the German Army, ensuring that the one organisation that could have stopped Hitler did nothing. All three had, in practical terms, helped the war effort perhaps as effectively as the card-carrying Nazis.

17

A New Beginning

When I had finished my articles for the *Sunday Express*, I returned to England. The country I got back to was not devastated like Berlin, but it was tired and grey, and in Coventry, Bristol, Birmingham and London I saw acres of rubble, boarded-up windows and peeling paint. Most people wore shabby clothes and managed on an austere but healthy diet, without much meat, butter or sugar.

The Milroys were relieved to see me; they had finished cleaning up my manuscript, which had become *Indian Route March*, and wanted me to check some details before sending it to a publisher.

'Now you can have a rest,' Dido said.

'I'd like to,' I explained, 'but I've got another idea.' And I told her I wanted to write a book about the people I had interviewed in Germany. The Milroys looked at each other and sighed but, largely thanks to their help, two years later a new book, *Follow My Leader*, was published and later reprinted in paperback as *The Mark of the Swastika*.

As I stayed again with the Medawars, the Milroys and other friends, I felt more and more at home in England. For ten years I had hoped for English citizenship, imagining I would feel quite different when it came. Then one day, a document arrived informing me that I was British, and to my surprise, I felt no different. It is only now, so many years later, that I realise becoming a British subject was one of the most important things that ever happened to me. In Germany, after 1933, I had felt a self-conscious stranger, a prisoner of the state. In England, I marvelled at the existence of a genuine opposition and admired the way people could debunk their own leaders and institutions, and agitate about things that did not directly concern them.

I looked forward to the future and, even though I wasn't really educated and didn't know how I was going to make a living, I felt confident that I would manage. I had good friends, good health and I trusted that my luck would hold and opportunities keep coming.

But before I started civilian life, I longed to see my parents again; I had not seen them for more than ten years. During the first few years of the war, I had lost all contact with my relatives, except for my youngest sister, Carla. As the baby of the family, she had been protected by her adoring parents from most of the unpleasant things happening in Nazi Germany. But when she was 12, she was made to stand in front of the whole class as an example of the 'inferior Jewish race'. She never quite got over this experience: it made her frightened of life and took away the self-confidence and optimism she had shown as a child. When she was 14, she was sent to a boarding school for refugees near Haslemere in Surrey, about an hour's drive from London. I was her only relative in England and managed to see her every two months or so.

I never worried about my other siblings – KV was working in a stockbroking firm in New York, Nina was married to Max Jacobson, a successful New York doctor and Peter had got a job in Long Island as a butler with a rich American family – but I worried desperately about my parents, and how my mother especially was coping with the terrifying conditions they must have been experiencing.

In 1938, when my father had at last realised the Nazi madness was not going to go away and war was inevitable, he had started to make plans for them to leave Germany. By then it was extremely difficult for Jews to get away. The authorities had deprived them of any rights and had stolen their money and property – the 'final solution' would take care of the rest. My father bought forged Finnish passports with some money he had prudently secreted, and with those my parents took a train to Finland.

Unfortunately, something was wrong with the passports and they were caught at the frontier. My father was charged with possessing a false passport and sentenced to nine months in jail, a sentence that saved their lives. While he was safely in prison, most of the remaining German Jews were rounded up and sent to concentration camps. The Nazis did not like to undermine the authority of the regular police. The Orpo in Potsdam must have known my father well, and they saved him from the Gestapo while he served time in their prison.

Soon after my father's release, KV obtained a visa for my parents to emigrate to America and sent them tickets to New York on a freighter sailing from Santa Margarita in Italy.

Waiting the night before in a tiny third-class hotel overlooking the harbour, they were awakened by the Carabinieri and arrested. France had surrendered and Mussolini had thrown in his lot with Germany and declared war on the Allies; as a result, all German Jews were to be sent back to Germany.

At the station, hundreds of scared-looking people were brutally shoved into cattle trucks. For some reason, which was never explained, my parents were singled out and put into a tiny guard's compartment. As the train left the station, a young SS man joined them. My father tried to make conversation, only to be rudely rebuffed. The guard stared fixedly out of the window, muttering, 'Stinking Jews,' under his breath.

The journey went on for several days, and all the time more Jews were forced into the already overcrowded trucks. Still, my parents were left with the surly SS man. At last, they reached the Alps, and when the train entered the Brenner Pass, they realised they would soon be back in Germany. As they came out of the long tunnel, my mother stared at the guard, wondering what sort of man could send innocent people to their doom. Suddenly she had a flash of recognition and said, 'Aren't you a ski instructor from Galtür?'

He turned on her furiously. 'How did you know I come from Galtür?'

She said, 'I think you were our guide in 1931 when our party of children was caught in that terrible snowstorm.'

'Perhaps,' he growled, 'but who are you?'

When they told him their names, he remembered the whole episode. 'Herr and Frau Hagen,' he said incredulously. 'But you were such good skiers: you can't be Jews. Without you, I would never have got those kids safely back to the hut.' After that, his manner towards my parents changed.

The next night, everyone from the train had to sleep on the bare concrete floor of a cellar in Augsburg. Towards morning, my parents were woken by the SS man. He beckoned them to follow him quietly. Outside in the street, he gave them 100 marks. 'Good luck,' he said. 'You are certainly going to need it.' And he quickly turned away.

My parents stood in the empty street without papers, ration cards, or the compulsory Star of David on their overcoats. At first, they wondered anxiously where they should go. Then the answer seemed obvious: they would go home.

When they reached Potsdam, good neighbours supplied them with food and necessities. They had left their remaining jewellery with friends for safekeeping, and those friends were now able, with influence and a certain amount of bribery, to arrange exit visas for them to travel to Japan, via Moscow and Vladivostock, on the Trans-Siberian railway. The journey was possible only because the German–Russian friendship treaty, signed in 1939, had not yet been broken. My parents embarked on the last boat to leave Japan before Pearl Harbor and made it to the safety of the United States, where my brother Peter was waiting for them on the dock in San Francisco.

By the time I arrived back in Britain, all my family were in the US and urged me to join them. I liked England and all my friends were there, but I was curious about America and thought I might be missing a great opportunity. I had some money left from the sales of *Arnhem Lift*, so I decided to go.

I found my parents happy; they had adjusted well to their new life in New York. My mother, who had always had servants, was very proud to do everything herself in the one-room apartment. I can still see her elegant figure standing at the electric cooker, slowly stirring vegetable soup with a spoon in her slim fingers, or proudly sewing on a button with long, slow movements.

A New Beginning

But this was not only the place where they cooked, ate, slept and received friends; it was also my father's workshop. When he arrived in America, he was 54, had only banking experience, and was as poor as a church mouse. He had to find another way to earn a living, something he could do into old age. He decided to learn radio repair and joined a school sponsored by the New York City Council. All the other pupils were in their late teens or early 20s, and called him Grandpa. After eighteen months, he passed with excellent marks and was asked to stay on as a permanent teacher, a post he held until 1946 when the soldiers were demobbed and needed their jobs back.

In America, after austere England, it was wonderful to indulge in eggs and bacon (still my favourite dish) and giant steaks, and to drive flashy cars. My family and their friends did everything to make my stay enjoyable, but my real attachment remained to England. I had decided to return, but first I wanted to see as much as possible of the United States. I did not feel like travelling alone, so began vaguely searching for someone to share the adventure with me.

At a party, I met a tall, blonde Norwegian girl who asked me if I had read certain recent books. 'No,' she maintains I answered, 'I have not read many books, I've only written a few.' We left the party together. She was called Anne Mie and had spent the war years in the US, working for the Office of War Information, broadcasting the news to Norway. When I met her she was 26, had been married twice to Americans and was recently divorced.

She had a job with a film company, taking stills of the Italian quarter of the Bronx, which was to be reconstructed in Hollywood. I was keen to see something of New York other than prosperous, streamlined Manhattan, so I went with her, carrying her heavy camera equipment. We soon became friends, and when her assignment was completed and she had to join the film company in Hollywood, I drove with her for three weeks across America. From Hollywood, we sailed to Norway on a freighter owned by Anne Mie's uncle.

Three years later, we got married. We had a house in London, close to Hampstead Heath, and another on the Oslo fjord, both generously given to us by Anne Mie's father. I tried to make a living writing books and translating others from German, but I was a slow writer and failed to write another bestseller. I could not earn enough to support our two homes and two daughters; I needed to find a different career.

Almost as soon as I arrived in England in 1936, I saw a lot of Lotte Reiniger, an old family friend I had known as long as I could remember. She had been the art teacher in the small private school in Potsdam; her husband, Carl Koch, had been the geography and history teacher. Lotte had been a genius at cutting out silhouettes and intricate illustrations with her scissors, and when she told my father how she longed to make a film with Carl, he agreed to finance the project and equip an animation studio in one of the school outhouses. There, between 1923 and 1926, she had created one of the first full-length animation films in the history of cinema, *The Adventures of Prince Achmed*. It is still frequently shown on TV and at special cinema performances with a live orchestra. After that, Lotte and Carl made several short animation films and a live action film with Jean Renoir, *The Pursuit of Happiness*. It was a disaster and ended the business relationship with my father, but Lotte found other sponsors for her films.

She emigrated to England at the same time as I did but for very different reasons: though pure Aryan, she and her husband could not abide the Nazis and everything they stood for. While Carl went to France to make films with Renoir, Lotte made publicity films for the British Government under John Grierson, head of the General Post Office Film Unit, later the Crown Film Unit.

After the war, our friendship continued and she often asked for my help in marketing her films. I spent more and more time doing that, and finally decided to have a go at producing them. Although we had promised Anne Mie's father never to mortgage our house, we used it to guarantee a loan to purchase two very old Debrie cameras and furnish a basic animation studio at the Abbey Arts Centre in New Barnet, where Lotte and Carl lived. With my old friend, Vivian Milroy, we founded Primrose Film Productions Ltd at the end of 1949.

A New Beginning

When we had completed more than thirty-five Lotte Reiniger short films, we thought it was time to branch out and make live action documentaries, mainly for German and American television. Lotte and I travelled together all over the world, to animation workshops and film festivals. She was a very amusing lecturer and the discussions after her talks usually ended with our expulsion by an impatient janitor. With a group of enthusiastic fans, we would find some late-night restaurant to eat and drink in until early morning. Lotte died in 1981, but I continued to distribute her films all over the world.

My writing career was not yet over, though. Early in 1992, the publisher Leo Cooper suggested *Arnhem Lift* should be revived for the fiftieth anniversary of the battle of Arnhem in September 1994. He urged me to write what I felt about it all, half a century on. Arnhem had been written about so much that I could not find anything original to say, but then a happy coincidence solved my problem.

At a dinner party given by a distant relation, I sat next to a distinguished-looking man about my own age. He was ex-major Winrich Behr, known to everyone as Teddy Bear. Now a prosperous retired businessman, he had been in charge of the German troops facing us in Arnhem. From him, I got enough material to add the German view of the battle. This new *Arnhem Lift* was published in 1993.

I did not mean to attend the official fiftieth anniversary of the battle. The idea of parading with hundreds of old veterans like myself, wearing rows of medals and red berets, did not appeal to me. But my publisher thought it was a good opportunity to promote the book. By the time he had persuaded me to attend, all the hotels and guest houses had been booked up and I was assigned to a billet in a barracks 31 miles outside Arnhem.

Quite by chance, a retired nurse, Ans Kremer, who was helping in the town hall with the housing of thousands of veterans and their families, had read my book and recognised my name on the list. She immediately wrote to me, offering to put me up in her house in Oosterbeek, where most of the fighting had taken place. I was delighted to accept her invitation, particularly as she offered to fetch me from Amsterdam Airport, 37 miles from Arnhem.

About ten days before my departure, I was surprised to get a telephone call from someone who called himself Prince Bernhard of the Netherlands. I thought it was a practical joke, but he went on speaking in perfect German, using the familiar 'du', and I realised it really was Prince Bernhard. He complimented me on my book and said he wished I had spoken to him before I wrote the last chapter, in which I blamed Field Marshal Montgomery for the disastrous failure of the airborne action. He said I had been too gentle. He and a Dutch major-general had tried to warn Montgomery that the 9th and 10th Panzer Divisions were refitting near Arnhem. Montgomery had refused to listen. The presence of so many tanks had already been confirmed by listening to signals from the German High Command; Montgomery had also ignored these and never passed on the information to the officers in charge of the Arnhem action. All this would indeed have been a valuable addition to the book.

But what Prince Bernhard was really telephoning for was to invite me to come to the palace when I arrived. Not thinking, I asked him for the address. He laughed. 'Oh, everybody knows where it is.' I explained that I was being met by Ans Kremer and her brother, and he told me to bring them, too. I did, and we all had tea in Prince Bernhard's study at the palace.

During the drive to Oosterbeek, Ans told me they were now almost certain they had met me fifty years ago when she would have been 12 and her brother 10 years old. They remembered the incredible moment when, after four years of German occupation, the sky above Oosterbeek was suddenly filled with hundreds of men and parachutes silently floating down. Mortars and shells started exploding all around and their parents had hustled them down into the cellar. On the third night there, they'd heard a persistent banging on the front door. Finally, Ans's father thought he had better open it and Ans crept up after him. In the dark they saw a man in a uniform they did not recognise calling out, '*Sprechen sie Deutsch?'* then in perfect German asking if any of the houses were occupied by Germans: only then did they realise this was an Allied soldier come to liberate them.

Now that she heard my voice, she thought I might have been that man. During the battle, we had spent five days in the Kremer house; we used their heavy Victorian furniture to barricade the windows and dug trenches in the front and back gardens. When we got to their house, Stationsweg No. 8, Ans fetched a tattered, leather-bound visitors' book and opened it to a stained and blotchy page on which I had written, above the signature 'Lewis'. 'I do hope and believe that the mess we made of your lovely house was worthwhile,' I said, 'and good luck for a happier future.'

The Hagen family skiing. The man standing in front on the right is very likely to be the SS guard who was previously their ski instructor, and who later saved their lives.

Louis' parents in their apartment in Florida.

Louis' parents in Florida, by their car.

Family photo (Louis sr, Louis jr, Anne Mie Hornemann and Vicky), California, 1947.

Anne Mie and Louis' wedding, 1950, 10 Ellsworthy Road, London.

The interior living-room view of 13 Highgate West Hill, where Louis and Anne Mie brought up their family. Louis lived there until he died.

Louis' family eating round the table. Louis, with two of his daughters, Caroline and Siri; Anne Mie, left, and cousin Unn, with their two Norwegian au pairs.

13 Highgate West Hill, exterior view.

Lotte Reinegar and a still from her film *The Adventures of Prince Achmed*. Louis formed Primrose Film Productions Ltd with Lotte; she started out as the children's private art teacher in Potsdam.

Louis and Vicky Hagen, Carl Koch and Lotte Reineger, Karl Vicktor Hagen and Schatel, Egypt, 1927.

Epilogue

In 1989, unbelievably, Russian and European communism fell apart. I was in Berlin on film business soon after the fall of the wall. At last, it was no longer difficult to cross the border between west and east, and a childhood friend drove me to Potsdam. Our big house on the Jungfernsee had become an eyesore. In the years after my parents left, it had been used as a youth hostel, then as a rest home for high-ranking party officials. Finally, it had housed Stasi computers and the water police who, only a few weeks earlier, had been chasing and shooting East Germans trying to escape to the west, 450m away on the other side of the lake. The two huge weeping willows that flanked the house had been cut down and a long corrugated iron building erected on the lawn leading to the lake. At one end stood the large crane used to lift on to dry dock the speed boats of the water police.

It was the same story at Carlshagen, my grandfather's summer residence in Potsdam. It was like a château built at the turn of the century, surrounded by a large park stretching down to the banks of the lake. During both world wars, it had been used as a convalescent home for injured army officers. When I got there, it had been turned into a children's hospital. I could still make out its former splendour – the marble fireplace in the hall, the plaster mouldings and oak panelling – but the exterior looked especially dilapidated and I was shocked to see an eight-storey block of flats on one side and an ugly sports centre on the other.

As for the family bank, my German friends had warned me that the entire business centre of Berlin had been razed to the ground during the bombing, but I still thought I might find some ruins of it, and eventually I did: an untidy plot covered with rubble.

Seeing the terrible changes in these familiar old places shocked me so much that it did not occur to me they would one day be returned to us. A year or so later, a law was passed by the German parliament stating that everyone who had been forced to sell their property during Nazi rule could claim it back, no matter who now owned or lived in it. Only then did I and the rest of the Hagen heirs, most of us by this point over 70, realise that in our old age we might become quite rich. At first we felt euphoria – until it became clear that actually getting the properties back would be a long and arduous process. Though Jewish claims were to take priority, we were competing with millions of Germans who had also had their property seized. Fortunately, my nephew, Dr Louis Hagen, an American citizen who had studied banking and law in Germany, devoted his entire free time to getting our property returned. He employed four lawyers and handled the endless red tape and correspondence with a ministry that was not only grossly overburdened and understaffed, but unsympathetic and sometimes simply bloody-minded.

By 1995, Louis and his team had, despite incredible difficulties, managed to return to us our Potsdam house and the Berlin bank plot. The house was eventually sold to Peter Block, the son of the architect who had built it, for 2.3 million marks (about £1 million at the time). Soon after the restitution law came into effect, the bank plot had been illegally sold at a knock-down price by the Bundesamt Zur Regelung Offener Vermögensfragen, the office for the settlement of property and wealth questions. It took years of negotiations, legal threats and direct approaches to senior officials to recover the 8.3 million marks the ministry had pocketed. Louis had even persuaded my cousin Lutte, who was nearly 80, to make an arduous journey with him to meet the federal finance minister in Bonn, to emphasise the urgency of settling the claims while the Hagen heirs were still alive.

Epilogue

The restitution of my grandparents' house in Potsdam is still outstanding and will take years and endless paperwork to resolve. But we have been lucky: those without legal help and financial resources have little chance of ever being compensated.

The effect of receiving the restitution money was more astonishing than winning the lottery, as no one had hoped or even thought of such riches. It changed all of our lives, some more than others. For my niece in the US, who lived on welfare as a single mother of two, it meant a decent flat, special schooling for her younger child and university for the elder boy. One cousin, who was also very poor and suffering in the last stages of cancer, could invite her children and grandchildren to New York, and enjoy arranging their inheritance before she died peacefully in a hospice. Three other cousins had come to England in their teens in the 1930s, when it was very difficult for a lone woman to have a career and become socially accepted. These girls had been brought up in prosperous, stable homes and found it difficult to cope abroad without the backing of friends or family. The oldest married a market gardener, who had become terminally ill; she could now afford a full-time nurse and to buy a house for her daughter. Another of them was a widow who let rooms to supplement her old-age pension. Now she could have alterations made to her council house to give her some privacy, and was able to send two of her grandchildren to university. The youngest of the three cousins, Irene, lived in a suburban house in north London and still longed for the luxurious way of life she had enjoyed as a girl in Germany. Now she is looking for a small, elegant flat in South Kensington and can help with the education of her two grandchildren.

For me, finally receiving the money meant no financial worries for me or my daughters, and I thought the family should celebrate with an 80th birthday party for me in Potsdam. Cecilienhof, the Crown Prince's summer residence, is now a luxury hotel and, being so close to our house, was the perfect place for it. At dinner there was champagne, many birthday speeches and much laughter.

The next day, I took my family to see our house. It was still unoccupied and guarded by the same caretaker, Herr Engel, who proudly told us he had been a political officer attached to the water police. He showed us where the machine guns and searchlights had been, and boasted that hardly any of those who had tried to swim the lake had made it. Noticing the shocked expressions on our faces, he assured us his job had been to tell people how fair and effective the communist government was, and to warn them of the aggressive intentions of the imperialist western neighbours. Although he seemed to have no qualms about switching his loyalty to the new capitalist regime, he told us how difficult it was to get by as a mere watchman. As an officer of the old regime, he'd had a lot of perks.

My wife and children had heard many stories of my youth, when the house was teeming with children, family and friends, sunning themselves on the terraces, so they were very disappointed by the ugly, bare building it had become. At Carlshagen, it was easier to imagine how my grandfather's house had once been, despite it now looking desolate and shabby, with doors and windows boarded up, signs of a fire on the upper floor and a tarpaulin covering the roof. The rear of the house, towards the lake, resembled Sleeping Beauty's castle, almost hidden by a lush wilderness we had to fight through to get a closer look.

The next day, we drove to Berlin to find Charlottenstrasse 92, where the bank had stood. In its place was a dreary shopping centre. But at the New National Gallery, a beautiful modern building designed by the Dutch architect Mies van der Rohe, could still be found paintings by Monet, Renoir and Manet, each bearing a brass plaque that read 'Donated by Carl Hagen, 1906'.

Louis by the Berlin Wall in 1990, soon after it came down in 1989, revisiting Potsdam for the first time.

The dilapidated house at Bertinistrasse, Potsdam, in views from the street and the water.

A family photo of Louis' 80th birthday, beside his grandfather's dilapidated house (seen below).

Louis, with Ans Kremer and her brother, visiting Prince Bernhard on Arnhem's fiftieth anniversary.

Ans Kremer's house, which Louis and his unit occupied.

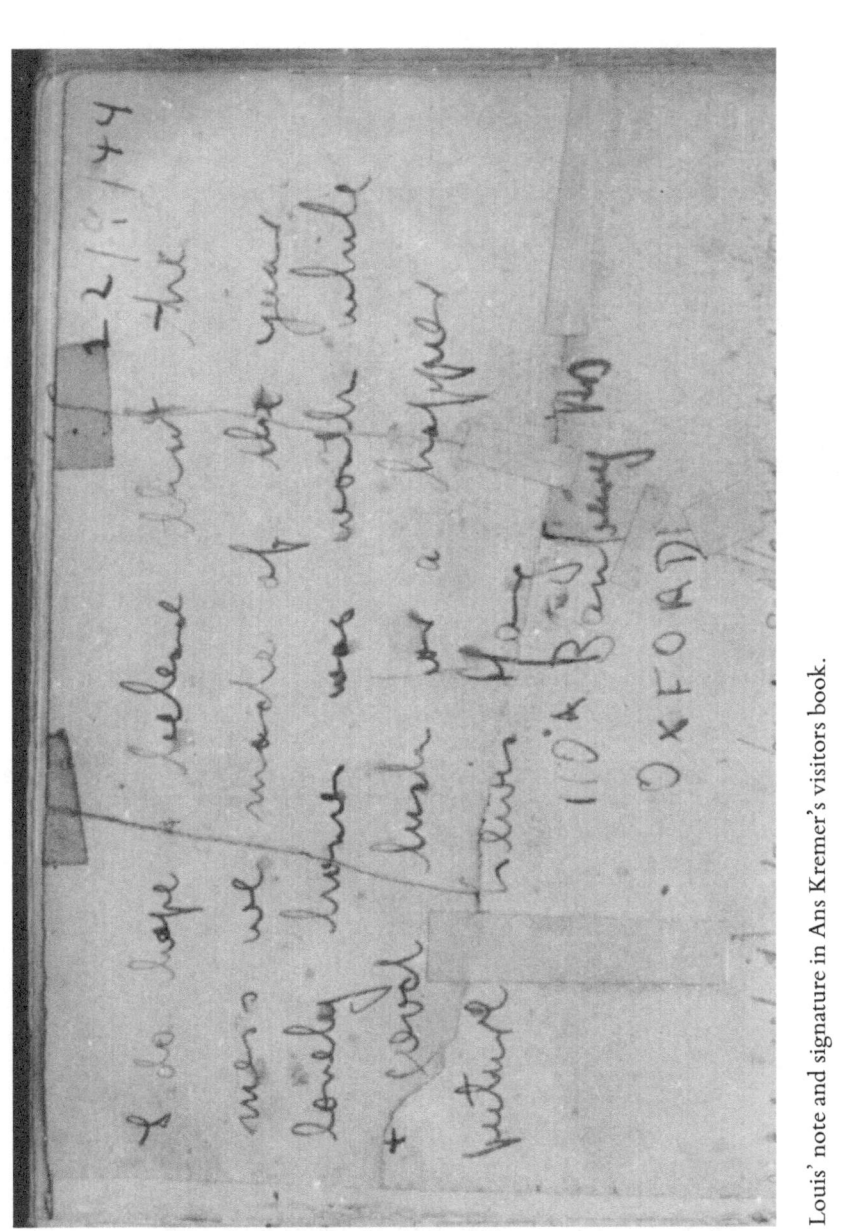

Louis' note and signature in Ans Kremer's visitors book.

'Saint Germain l'Auxerrois', Claude Monet (1840–1926).

'Der Nachmittag der Kinder in Wargemont',
Pierre-Auguste Renoir (1841–1919).

Anne Mie and Louis in their later years, in the garden at 13 Highgate West Hill.

ARNHEM LIFT

A GERMAN JEW IN THE GLIDER PILOT REGIMENT

LOUIS HAGEN

To all who fell at Arnhem – Allied and German

'I crawled right under the brushwood and saw and heard the bullets splashing the ground and hitting the branches and tree stumps all round me. I was sure this was going to be the end and kicked myself for doing such an idiotic thing; trying to take a strong German position on my own. I swore that if ever I got out of this hopeless position I would never again be such a bloody fool. I lay completely still, bullets whizzing about me. I wondered if I wanted to pray; that is what everybody is supposed to do in a position like this; but I just did not feel like it, and to calm and steady myself I watched a colony of ants go about their well-planned and systematic business.'

The action recorded here took place between 17 September and 25 September 1944. *Arnhem Lift* was first published in January 1945 (without the author's knowledge!). Not only is it an extraordinary story, dealing as it does with a German Jewish refugee who ends up flying a Horsa into the disaster, it is also believed to be the very first book published about the battle – to the fury of the military and to great public acclaim. In 1950, Louis Hagen married Anne Mie, a Norwegian artist, with whom he had two daughters, Siri and Caroline. Dividing his time between London and Norway, Louis Hagen established Primrose Film Productions and went on to create 25 children's films. He returned to Arnhem twice, first in 1948 to show his fiancée where he had fought, and again in 1994 for the fiftieth anniversary. He had not planned to attend as he felt 'the idea of parading with hundreds of old veterans like myself wearing rows of medals and red berets did not appeal'. Louis Hagen died on 17 August 2000 at the age of 84. He rests at Asker in Oslo, Norway.

Acknowledgements

With thanks to Jean Medawar, my oldest friend, who first taught me to speak English, and now helped me to write it.

I want to thank my old friend Vivian Milroy for his patient help and detailed research into the strategies of the Battle of Arnhem.

I am grateful to my former 'enemy', Major Winrich Behr, for telling me his experiences at Arnhem and for letting me use them in this book.

Louis Hagen, 1993

Prefatory Note to the First Edition

When the author of this book arrived home on leave, after fighting right through the Arnhem action, everybody wanted to hear his story. After telling it several times, he began to find the repetition irksome, so he spent the rest of his leave writing it all down, while the events were still vivid in his mind. Any more friends who asked him for the story would get a type-written document! That is his explanation of how it came to be written.

Then someone suggested he should publish it. A copy arrived on my desk. After glancing through a few pages, I settled down to an absorbed reading of what I found to be a remarkable piece of reporting. Other urgent matters were left aside, and willy nilly I had to read on to the end. I believe that other readers will find it equally compelling.

This young soldier had no public in mind, beyond a few personal friends, when he set down his adventures. This may account for the intimate quality of his writing. But there can be no doubt he has a flair for picking out those details and moments which we all want to hear about – the touches of unexpected realism which help us to visualise and live over again the incredible heroic episode of Arnhem.

I have struck nothing quite like this diary for giving the actual feel and flavour of modern war at its most spectacular. This is the story of one man's battle. It doesn't purport to describe the action as a whole. It gives instead a series of ultra-vivid images and experiences. Like real life, it is inconsequent and surprising. It is also straightforward and free from egoism. In spite of the grim nature of the ordeal, the author seems to have come through with a sense of elation rather than with the despair and nervous exhaustion which the First World War seems to have produced in most of those who have written about it. This resilience gives an effect of balance and composure to the story, which is very remarkable considering the hectic and often horrible things that happen throughout it.

C.M., 1945

Foreword to the Second Edition

This is the story of the First British Airborne Division's great fight to hold the Arnhem Bridge as seen and experienced by a glider pilot.

It is my purpose here to paint very briefly the bigger picture of the Airborne Operations in Holland, a picture which only the gallant survivors were to learn from me when they returned to Nijmegen after their withdrawal from the Arnhem perimeter.

The Second Army was faced with increasing opposition by re-formed German Battle Groups, in difficult country and with a series of water obstacles barring the way into Germany. Winter was approaching. The First British Airborne Corps (First British Airborne Division, 82nd US Airborne Division and 101st US Airborne Division) were ordered to seize and capture a corridor over 40 miles long which included the great bridges at Grave (River Maas), Nijmegen (River Waal) and Arnhem (Neder Rijn). The Second Army was to drive through this corridor and debouch into the North German plains, thereby turning the defences of the Rhine.

The Airborne Operation was successful in capturing the corridor and bridges, except for the essential gap between Nijmegen and Arnhem bridges. Owing primarily to the almost uncanny recovery by the German Army which had just suffered defeat in Normandy and on the Seine, and bad weather for air operations after the first two days, the Second Army were unable to reach Arnhem Bridge in time to achieve the complete breakthrough and the relief of the hard-pressed First Airborne Division.

This Division's outstanding action and the successful operations of the two fine American Airborne Divisions, though failing in their final object of passing the Second Army through into Germany, ensured the retention and consolidation of the high ground at Nijmegen, pushed the Second Army, with incredibly few casualties, through over the 40 miles of difficult country and over two great rivers, and provided the springboard from which 21 Army Group launched its final assault on Germany early the next year.

Without the First Airborne Division's heroic stand at Arnhem, which protected the northern flank of the battle and contained considerable German reserves, no such results would have been possible.

Lieutenant-General Sir Frederick A.M. Browning KBE, CB, DSO
October 1952

1

About the Author

When I fought in the Battle of Arnhem, I was a lighthearted young German who had been born into a wealthy Jewish family of bankers. I had been brought up by liberal and devoted parents in idyllic surroundings near Potsdam, on the outskirts of Berlin, a lovely small town where the Prussian royal families had for centuries spent their summers in elegant and luxurious houses on the shores of Lake Jungfernsee.

Now I am an Englishman nearing 80 [Louis is writing this in 1993], still lighthearted, married, with two children and two grandchildren, living comfortably in Highgate, 6 miles from the centre of London.

This book tells the story of the events that were partly responsible for my transformation.

When I was a boy I was so full of the physical joys of life that I learnt very little at school. I was more interested in 'taking dares'. I was dared to come into class with a false beard, or riding on a horse. I was dared to eat a live frog. I did all these things, and my reports were very bad. My parents decided it was a waste of time to keep me at school. One day, when I was about 16 years old, I heard my father say to my mother, 'There's no point in torturing the boy any longer. We'd better enrol him as an apprentice engineer.' So I left school and went to start at the bottom in the BMW works, where my father and grandfather were on the board of directors. I enjoyed this new working environment and I learnt a lot, without losing my high spirits.

One day I stupidly wrote a vulgar postcard about Hitler's Brownshirts to my sister Nina.[1] She left it lying around and it was picked up by one of the maids. This maid had been stealing my mother's jewellery and was about to be sacked. She threatened to take the card to her boyfriend – who was one of Hitler's storm troopers – unless my mother withdrew the accusation of stealing.

This was in 1934 and neither of my parents had any idea how seriously the new National Socialist (Nazi) authorities would treat the incident. So the maid was sacked; she went to her boyfriend, and soon after that storm troopers appeared at the BMW factory to arrest me. I was taken under guard and shut up in Torgau, an old castle which had been turned into a concentration camp. I was with a mixed group of men who were considered 'racially inferior' like Jews or gypsies, or were thought to be politically dangerous, like communists and Freemasons. The camp was run by Nazi storm troopers who enjoyed humiliating and torturing elderly Jews, or boys like me who had been brought up with more money and education then they had enjoyed. I was kicked off my straw palliasse in the middle of the night and made to crawl about naked on all fours while I was beaten for the amusement of the drunken guards. We had to empty latrines with our bare hands and carry the contents away in heavy iron containers that cut into the skin of our palms.

I was comforted by a fat, middle-aged communist called Wolfgang who became my friend. One day, while I was playing chess with Wolfgang, I noticed a group of men crowding round the window facing the courtyard. We got up to see what they were looking at but before I could get there Wolfgang pulled me back: 'There's nothing there, Büdi. Let's get on with the game.' But I insisted on seeing what was happening: if ever I got out, I wanted to be able to tell people what was going on.

What was happening was the cruellest, most shocking thing I have ever seen. It was a very hot, sunny day; a group of SA men in shirtsleeves

---

1 The note read, 'Toilet paper is now forbidden, so there are even more Brownshirts.'

were standing round the farmyard pond where there were usually a few ducks swimming and a couple of pigs cooling themselves in the mud. There were neither ducks nor pigs in the pond, but instead four prisoners splashing around, entirely covered in mud, moving as if in slow motion because of its dragging weight. They were trying to crawl out of the pond but whenever they reached the edge, the SA men kicked them back in, laughing and shouting. I could not go on watching and turned away. Later I learnt that none of them survived.

When the guards could think of nothing better to do, they gave each of us a bucket of water, then chased us round the courtyard. If we spilt any, they beat us. I was young and very fit and I managed, but some of the older men were soon exhausted in the blistering heat and collapsed. They were then kicked and beaten until they got up and started running again. The weakest were chased into the muddy pond to 'cool down' before, covered in mud, they had to start running again.[2]

After six weeks of this I was rescued. One day, standing to attention in the courtyard, I saw the gates open. A large black Mercedes drove through, flying the swastika flag. In it was the father of one of my friends at school – a judge.[3] I heard him ask for me. I was escorted to the guard-house where the commandant informed me that, if I ever revealed what was going on in the castle, 'We will get you, wherever you are, and bring you back, and then you will never get out'.

After my concentration camp experience my parents realised how dangerous it was for their five children to stay in Germany; my father thought it unnecessary, however, to make preparations for himself and my mother to leave. He used to say 'This Hitler business is too crazy; it won't last'. He thought the Nazis would not do anything to

---

2 The description of murder in the duck pond is taken from Louis Hagen's unpublished biography.
3 This was the father of Claus Furhmann, a boy Louis had befriended at school. Apparently the relationship then was mutually beneficial, Claus helping Louis with his school work, Louis acting as Claus's minder. Though Louis could not have known at that time quite how beneficial the friendship would prove to be.

him because he had been decorated when a naval officer in the First World War and was head of one of the oldest and most respected banking families.

Arrangements were made to get me out of Germany as quickly as possible. It was not easy, in spite of my family's many connections abroad, because at this time thousands of Jews were also desperately trying to leave. One country after another tried to control the flood of refugees from Germany; they all had problems with unemployment and were in a severe economic depression. It was almost eighteen months before a business friend of my father, Sir Andrew McFadyean, arranged for me to emigrate to England in January 1936.

Sir Andrew, chairman of the Liberal Party, was adept at getting German refugees into England: among others, he rescued the Hamburg banker, Sigmund Warburg, who founded the well-known London bank, S.G. Warburg. Sir Andrew managed to get me a job with the Pressed Steel Company at Cowley, near Oxford, which made car bodies for Austin, Morris and other British car manufacturers. I found that my apprenticeship at BMW helped me a great deal.

The job lasted almost three years. Then one day I was summoned to see Mr Müller, the general manager. He told me with some embarrassment that no foreigners were now allowed to work in factories engaged in war work; he had no option but to dismiss me. He assured me that he deeply regretted this because he had only had good reports about my work. He asked me what my private circumstances were and what I would do now. I told him that I had no family in England and no private money because my family were not allowed to send money out of Germany, and that my permit was valid only for the Pressed Steel Company. Mr Müller arranged for my salary of £3 per week to continue and promised that he would try to help me.

Ironically, only three years earlier the managing director of BMW, Herr Popp, had called me to his office and told me that one of the most embarrassing things he had ever had to do was to dismiss me – the Ministry of War would not in future permit 'non-Aryans' to work in factories producing armaments.

About the Author

Mr Müller soon arranged for me to be transferred to the sales and service department of a subsidiary, Prestcold Refrigerators, in London. But when war broke out even this job folded because I was then officially classified as an 'Enemy Alien'.

Now I was jobless I kept myself busy as a carpenter, building stage sets for small experimental and club theatres. I got hardly any pay and had to give up my flat, so I stayed with various friends and often slept on the stage wrapped in one of the curtains. I actually enjoyed this free and easy life.

I was of course elated at the prospect of the Nazis' downfall and like most of my friends, I went to one of the many recruiting offices to volunteer for the army. An officer there told me that I would be informed when and where I was to join His Majesty's Forces. As all the others who had volunteered were awaiting their call-up papers, I was not worried when I did not hear anything for a long time.

At last I did get a notice, forwarded by Sir Andrew, to appear at a tribunal to decide whether I was an anti-Nazi or a dangerous enemy alien, to be interned for the duration of the war. Sir Andrew offered to appear at this tribunal, but I told him I did not think this was necessary as my case was so clear – I was of Jewish extraction and had been imprisoned in a Nazi concentration camp.

When I appeared at the tribunal, the chairman, before he even asked me to sit down, barked at me, 'How is it you arrived here in your own car?[4] You were supposed to report regularly to the police, not to be away from your home overnight and not to travel more than five miles from your registered address. You complied with none of these regulations and the police were only able to trace you through Sir Andrew McFadyean.' He went on dressing me down for several minutes. Luckily for me Sir Andrew had decided to attend the tribunal and now offered to testify on my behalf. He said that he had known me and my family for over fifteen years, that I had no close family in this country and had only English friends, and therefore had not realised that aliens had been supposed to register with the police when the war started. His testimony

---

4 Enemy aliens were forbidden to own or drive cars.

saved me from internment – probably in Australia or Canada – and after a further strong reprimand from the chairman I was free, until my call-up, to return to my jolly life in the theatre.

It did not last long, however. A few days later I was arrested as a deserter because I had not reported for military service. I was stuck in a cell in Notting Hill police station. Without a fixed address, I had of course never received the call-up notice. My guardian angel, Sir Andrew, again saved me and, after I had spent two nights in the cells, I was escorted by Military Police to the Pioneer Corps training camp at Westward Ho! near Bideford in Devon. The other recruits cheered when I answered my name at the next morning's roll-call. For weeks they had got used to the dead silence whenever my name had been called.

I soon caught up with the military training, marching, polishing boots and brasses, laying out equipment, making my bed accurate to the millimetre and presenting arms with broomsticks. We were not issued with rifles because there were not enough for the huge number of recruits – and possibly because we were not yet entirely trusted. Nor could we yet hope to become commissioned or non-commissioned officers. By the time the British Expeditionary Force was being evacuated across the Channel from Dunkirk, we so-called Pioneers were stationed along the Devon coast defending our newly adopted country with sticks and clubs. But it was a lovely hot summer and we swam and sunned ourselves whether we were on guard duty or not.

When our initial training was over we were sent to dig latrines in remote open country where airfields and army depots were to be built. Although the work was hard and we were often very wet and cold, we were in good spirits. At last, and at least, we were doing something to defeat German fascism. Most of us came from an educated Jewish middle class, so there was none of the usual barrack-room obscenity. Everyone was most civilised and one often heard conversations like this: '*Nehmen Sie bitten den Eimer, Herr Doktor. Er ist aber sehr schwer*'; '*Vielen Dank, Herr Professor. Ich schaffe es schon.*' ('Please take the bucket, Doctor, but it is very heavy'; 'Many thanks, Professor, I think I can manage.')

About the Author

We were a strange collection — professionals, academics and a few musicians and writers — and our motto was lighthearted: 'Vy vorry? Ve vill vin ze var'.

Slowly the War Office began to trust us (as far as I know there was not a single case of espionage among the 15,000 or so 'enemy aliens' in the British forces during the entire war), and one by one other units and regiments became open to us. I soon found out that I was not the type for a military career and decided to volunteer for anything that would relieve the boredom and might teach me something or offer me a new experience.

First I volunteered for the Army Fire Brigade, then for the Ordnance Corps, then the Royal Electrical and Mechanical Engineers (REME), the Artillery Assault Corps and many others. I was invariably accepted; I was the right age and physique and I could read and write. The many journeys to and from these selection boards gave me plenty of opportunity to get lost on the way so that I could visit my friends in London and Oxford.

One of the very last regiments that was prepared to accept 'enemy aliens' was the Army Air Corps' Glider Pilot Regiment, which was officially formed on 24 February 1942. Of course I immediately volunteered for this: to become a pilot was the ultimate goal, not only for me, but also for thousands of other soldiers of all ranks.

I have never been to an interview in such a state of excitement. I felt that if I were accepted I would be the luckiest man alive. I still remember how I disliked being treated as a third-class citizen in Germany and a foreigner in Britain. Becoming a British pilot would be a miraculous change.

When I stood in front of the selection board I was trembling, and when the chairman, a colonel asked me, 'Where do you come from?' I became speechless. If I said 'Potsdam' would I have to go into a lengthy and complicated explanation? If I said 'Oxford' would they call me a liar?

'Relax, man, and answer. We're not going to bite you,' the colonel said, smiling. His easy manner brought me back to my senses. So I said 'Potsdam, Sir' and after a moment of surprise he asked me to explain. I did so and then he said, 'My second question would have been, "Why did you volunteer?" but I see now that is superfluous.'

Two weeks later (it was now late in 1943) I was queuing up for my flying gear at the RAF airfield at Denham, just outside London. Flying did not come easily to me and it was rumoured that unless we were ready to go solo after seven hours' tuition in the air we were 'out' and would have to return to our army units. Luckily I had a kind and understanding instructor.

We trained in the famous de Havilland Tiger Moth, a beautiful open two-seater biplane designed in 1923. Day after day flying lessons went on, some lasting for only a quarter of an hour. Every minute in the air was recorded in my log-book, but as the seven hours crept nearer, I could not imagine being all by myself. When the flight book recorded eight hours and fifteen minutes my instructor jumped out of his seat and said, 'It's all yours. I don't dare to watch. I'm going to have a cup of tea in the canteen.'

I took off and stopped thinking of anything else except flying. It was suddenly a wonderful feeling to be all by myself in the endless sky. I flew the prescribed circuit inspired by the joy of flying alone. My landing was OK – not perfect, but good enough not to damage the undercarriage. I was over the first hurdle.

But we could not get our 'wings' until we had passed tests on the theory of navigation, aircraft recognition and the principles of flight. I did not find this easy and had to spend all my spare time revising. I was continually interrupted by kit and rifle inspections, interminable drill parades and random roll-calls organised by a sadistic army sergeant-major who was determined that we should not succumb to the lax RAF ways.

One day I was studying hard in my Nissen hut when the door was flung open and the sergeant-major barked out, 'You lazy lout! Why aren't you on parade?' I explained that I had been excused parades so that I could get ready for the next test. 'You've got all night to do that,' he said 'I wish I had,' I replied, 'but your stupid regulations mean we have to be in bed with lights out at ten o'clock.' He stared at me, then roared, 'Now you're insulting a superior officer. You'll regret this, young man. Report to Company Office, 1830 hours.' I realised that the sergeant-major had had it in for me for some time. And the duty officer sided with him – without even asking for my side of the story. 'We don't need people like you,' he said coldly. 'Hand in your kit. Get your travelling papers and return to your unit. Dismiss!'

I went to say goodbye to my instructor. All he said was, 'What nonsense! We can't lose good chaps like you. I'll talk to the Squadron Leader.' I spent a very long and tense ten minutes. Then he came back, patted me on the shoulder and said, 'Go back to the stores and collect your gear before it's issued to someone else.' I could have hugged him, but made do with a half-choked 'Thank you, Sir'.

Before I got my wings as a second pilot, a new directive announced that all Germans and Austrians in front-line units had to be issued with new identity papers, with English-sounding names and British places of birth and next of kin. The reason for this was that, if any of us were taken prisoner and the Germans discovered our nationality, we would be tortured for information and then shot as traitors.

I went home to Oxford and there my closest friends, Jean and Peter Medawar,[5] suggested the name 'Lewis Haig' – a name I kept until a year or two after the war. The army thought it a great name because of the even more famous whisky. I was registered as born in Oxford and the Medawars were listed as my next of kin.

I never got my first-class glider pilot wings because, on 17 September 1944, we were ordered into action. Our destination was Arnhem, a small Dutch town on the Lower Rhine. What happened then is described from Chapter three onwards.

---

5 Later Sir Peter Medawar, OM, CH, CBE (1915–1987), the distinguished medical scientist and Nobel Laureate.

2

The Background

After five long years of a land war against Germany, which had flowed from Poland to the Low Countries, Scandinavia and France, North Africa, Greece and Italy, the focus moved back to Britain in 1944. Here, in June of that year, the joint American/British forces mounted the greatest sea/air invasion in history. Three million men, half a million motor vehicles and tanks and three complete mobile harbours were ferried or flown or towed over 100 miles of English Channel.

It was not easy. In three days the Allies lost over 3,000 troops – but a bridgehead 2 miles wide and 3 miles deep was established in Normandy. General Dwight Eisenhower, the Allied Supreme Commander, now commanded two and a quarter million men, who, in the next few weeks, were to advance through France against the tough and battle-hardened German troops.

As they advanced, the problems of supply grew. Half a million motor vehicles and armour needed a vast amount of petrol, oil and ammunition; and two and a quarter million men needed a constant supply of food, ammunition and equipment. Everything had to be shipped over the Channel either to the temporary Mulberry harbours, or on to bare beaches and then on to lorries for delivery to the forward units, the Allied bombing having put the French railways out of action. The supply lines soon stretched to the point where the advancing Allied troops had to slow down: they finally came to a halt at the beginning of September near the Dutch/Belgian border.

Now there was a new danger. A large army just standing still might well encourage the enemy to counter-attack: then, with winter coming on, the battle could develop into the bloody and inconclusive stalemate of trench warfare.

In order to break the deadlock, Field Marshal Sir Bernard Montgomery, commanding 21 Army Group, the northern 'prong' of the Allied advance, proposed early in September 1944, using airborne troops to make a daring leap over the German defences and open up the way for a mass advance by the Allied armour into Germany. British and American parachutists and glider-borne troops would land in strength behind the German lines in the Netherlands, capture bridges over the Meuse, Waal and Lower Rhine, and hold them while Lieutenant-General Miles Dempsey's British Second Army, led by XXX Corps, poured through the Netherlands into Germany. Montgomery persuaded Eisenhower to support his plan, which went ahead under the code name 'Market Garden'.

The details of the plan, as evolved by Montgomery with Eisenhower's approval, called for the following:

> Parachutists of US 101st Airborne Division would capture the canal bridges at Son and Veghel.
>
> US 82nd Airborne Division would take the bridges over the River Maas at Grave and the River Waal at Nijmegen.
>
> The bridge over the Lower Rhine at Arnhem – the farthest bridge – would be the responsibility of British 1st Airborne Division. Simultaneously with the air landings, Lieutenant-General Brian Horrocks' XXX Corps would strike out northwards from the Neerpelt bridgehead and relieve the airborne troops in turn once they had captured the various bridges.

If successful the operation would establish an extended supply line as far as the last bridge at Arnhem. The great natural barrier, the Rhine, would have been crossed, and the Ruhr and the rest of Germany would lie open to the Second Army with no more natural obstacles in the way.

The details of that part of the plan which concerned Arnhem were worked out by Lieutenant-General Frederick ('Boy') Browning and Major-General Roy Urquhart, respectively the commanders of the British 1 Airborne Corps and 1st Airborne Division. (The two American airborne divisions were to come under Browning's 1 Airborne Corps for 'Market Garden'.) One factor that gave them some concern was that the RAF had insisted on the drops being made well outside the town, in order to avoid the anti-aircraft guns that were protecting Deelen airfield. Urquhart was unhappy about this, but such was the urgency that it was by then too late to refer the plan back to Montgomery.

The 1st Airborne Division plan called for 1 Parachute Brigade to capture and hold, in order, the Arnhem railway bridge, west of the town, a pontoon bridge about 2½ miles east of that, and, finally, the crucial road bridge 600 yards further to the east. 1 Airlanding Brigade would protect the landing zones until the second wave, which would include 4 Parachute Brigade, landed on the second day; they would then move east to link up with 1 Parachute Brigade, who would be holding the bridges. 1 Airlanding Brigade would then move to the west and form a defensive wall. Finally, Polish 1 Independent Parachute Brigade would land south of the river on the third day and march over the bridge to join up with 1 Parachute Brigade.

The bulk of the division – some 10,000 men in all – with their weapons and equipment, were to be landed in three stages on 17, 18 and 19 September on two landing zones and two drop zones situated from 6 to 10 miles to the north-east of Arnhem, in the neighbourhood of the hamlets of Wolfheze and Heelsum.

A number of supply and reinforcement drops were also planned, to be activated as the situation demanded. If all went well the Division would be relieved by Horrocks' XXX Corps within two days. Even under the most unfavourable circumstances they would expect the operation to be over in four days. The path would then be clear for the bulk of Second Army to pour up the corridor, occupy the Netherlands and turn east into the industrial Ruhr.

3

Arnhem Lift

Anyone who went to Arnhem could have told this kind of story. Mine is for the friends and relations of the men who did not come back …

Monday

We knew it was coming off this time as the first glider lift had left on Saturday morning. We were waiting in the mess for the tug pilots to return and give us the gen. All seemed well. They had found the LZ (landing zone) quite easily, with no flak to complain about and, as far as they could see, there was no ground resistance. We were all happy and confident about our lift on Monday morning – this time we knew that it was not going to be cancelled, as once a large-scale airborne operation like this has started nothing can interfere with its planned development.

We were lucky to be one of the first planes to take off on Monday morning. I was second pilot – the first pilot, Mac, was a typical Glaswegian. Our load consisted of one jeep with a trailer, both loaded with petrol, and three chaps of the Parachute Brigade. They rather resented coming with us because to them a glider is an unknown quantity and an extremely dangerous way of travelling. They feel much safer jumping with the other lads.

It is hard physical labour flying a glider in the slipstream of another aircraft, but our tug pilot was very skilful in avoiding the hundreds of other planes making for the Netherlands. He had to fly completely out of formation and at the wrong altitude to achieve that, but we encouraged him and praised him all the way.

Map reading did not seem much use to us, once we got over the sea and were approaching the Netherlands, as the Germans had flooded all the large islands in the Rhine delta and great stretches of the country itself were under water. Dry land was slowly emerging and I thought it time to check up on our position.

'Hello Tug ... Matchbox here ... How many minutes before we reach the LZ? Over.'

'Hello Matchbox ... Tug here ... Another fifteen minutes and it's all yours ... Can you pinpoint your position? ... Over.'

'Thank you, Tug ... I should say we were just crossing the first of the three arms of the Rhine delta ... Please confirm this ... Over to you.'

'You are correct ... Now only two more river crossings and you should see your LZ.'

From then till we landed it was essential that I should not lose our position for a moment. I continually checked from my map to the ground and peered searchingly forward for the first signs of our objective. Some flak came up at us, but it was very light. Our ideal pilot still kept us away from the rabble and out of any slipstream. I made a mental note to buy him a pint when we got back.

Soon I recognised the Lower Rhine, and a moment later could see our LZ – two small squares of wooded land pieced together at one corner only. Our landing was to be just where the woods joined together. It looked exactly like the photographs they had shown us at the briefing. I never imagined they could possibly look so much alike.

This was the moment to cast off:

'Hello Tug ... We are getting ready to cast off now ... Thanks for the wizard ride.'

'Best of luck, Matchbox ... See you soon.'

'OK Tug ... Same to you.'

Mac pulls the lever which releases the cables from our wings and we are in free flight. The tug banks off to the right as Mac pulls up our nose to gain height and reduce our flying speed. As we settle to our normal gliding speed, all the noise dies away and it seems unbelievably peaceful and calm in our cockpit. We can't think of it as other than one of our many mass landing exercises. We are now slowly losing height, and as we cross the river we can clearly see the bridge at Arnhem which is our ultimate objective.

We are nearly there now. We turn to starboard with half flaps down and our gliding angle steepens suddenly. Another 15 degrees to starboard and we are just about over our landing area. Full flaps down and our nose is now pointing directly to the ground, the flaps keeping our speed constant and just above stalling speed. Someone cuts in from the right and we veer off a little, and then, just before we hit the ground, pull out level. We lift gently over a hedge and then touch down firmly. Brakes full on ... a slow skid to port ... a perfect landing.

We sit there for a moment, looking pleased with ourselves, when the crackle of distant machine guns and the whistle of some nearer shots, which are obviously meant for us, remind us forcibly that this time we are not on an exercise. We leap out and get into the tail unit, which we had to remove before the jeep and trailer could be got out.

Mac and I start on the heavy bolts inside the tail, eight of them, and they have to be synchronised. Meanwhile, two of the parachutists loosen the shackles on the jeep and trailer, and the third one begins to cut the control wires. Mac and I are sweating like pigs. We have to work together and reach the same stage of the operation at the same time. We have to be quick. They are still sniping at us and we are completely helpless and exposed. Safety wire cut ... backwards and forwards with the release lever ... one by one the bolts come out ... not so hard really ... pretty much like the drill on the station ... I'm stuck now ... the bastards are getting more and more difficult ... I'm so terribly hot. I send one of the parachutists to stick the trestle under the body and he shouts that he has done so ... we get to the last two bolts ... must be completely together now ... mine is quite loose ... ready Mac? ... right ... go! ...

why the hell doesn't the tail fall off? ... we've done everything just like the practice ... we bang from the inside but it is stuck fast. I jump out to look and discover that the bloody fool of a parachutist has stuck the trestle under the tail itself. ... I kick it away, and with a terrific crash the whole tail fuselage breaks off and falls over on the trestle ... it's in the way still ... we all get our shoulders under it and heave to the left ... but the trestle is jammed in the fuselage now and embedded in the ground ... it's a hell of a job ... eventually we manage it, sweating, cursing and using all our strength ...

The two runners slipped out and fitted beautifully, and the paratroop driver drove straight out of the glider. As we jumped on the jeep and drove off, we noticed just on the right a cross and grave of one of yesterday's glider pilots. We had been surprisingly lucky. Most of the first and second lifts had taken a lot of punishment before they even reached the ground. We had not really done so badly because now that we had a chance to look around, everywhere we saw groups of men cursing, sweating and heaving to get the tails off their gliders. Some were even using saws and axes, and when we looked at our watches we found that we had done the job in twenty minutes, which made us feel very pleased with ourselves. To get to our first RV we had to follow a narrow sandy lane through low brushwood, small fields and single rows of trees. Everywhere we saw gliders; in the fields, some even on the trees, there were an odd wing wedged between two big branches of an oak, a tail unit sticking right up in the air and pieces of gliders distributed everywhere. We passed a large meadow with gliders parked in a more orderly fashion; obviously this was the real landing zone of Sunday's lift. We joined more and more jeeps and trailers, all filing to their various RVs. Ours was not so difficult as it was Wolfheze station, and from there to the lunatic asylum. It was a tiny station and its main features were the cross-roads running over the lines and parallel to the railway. Here was a terrific assembly of jeeps, trailers, light artillery and groups of parachutists. Red Cross jeeps with stretchers bearing casualties were passing through, nurses and men vainly trying to repair the water system in between this confusion. All the while we were sniped at; sometimes a mortar would go over and everyone seemed to disappear, but after a few seconds the confusion returned. Everyone was spreading out maps and asking everyone if they had seen or heard of their respective units.

In the crowd I suddenly spotted a glider pilot of a different squadron whom I had been hoping to meet for months. He had taken unfair advantage of me one day when I was about to go on leave. The urge to get away, once a leave pass is in your pocket, is so strong that any sacrifice is temporarily justified. This chap had dashed up to me when the car was packed up and ready to take us to the station, and asked me for the loan of a couple of quid, promising to return it by post, as we were unlikely to meet again. I never heard of him again and was very indignant, as he knew that I had no means of getting hold of him. I thought it was a dirty trick. There he was in the middle of the cross-roads at Wolfheze, and it all came back to me, and besides I had only ten shillings and no Dutch money on me. Sniping or no sniping, I started to dun him. He protested that he had very little cash on him, and an animated financial discussion proceeded. My troop shouted at me to come on as I had all the maps; they added that I could continue my argument when I got back to Whitechapel. We suddenly realised how idiotic we must look and both burst out laughing. He pressed a few guilders into my hand and I joined my troop again. We glider pilots were supposed to remain with the units we had taken over until further orders, and so we arrived eventually at the main building of the lunatic asylum.

There were scores of giggling and rather frightened nurses who screamed and scattered every time any kind of gun report was heard. The inmates had been moved out of hospital, and it was now filled to capacity with civilian casualties. Soon the blokes found out that I could talk to the civilians, and I was dragged from one group of nurses and paratroopers to another, translating. It was rather hard to tell who were the nurses and who were the inmates helping them. All were concerned to know where Prince Bernhard was and if their Queen had reached Dutch soil. They wanted to know which places had been liberated and if there would be any more bombing. It was quite a crazy atmosphere with our chaps, of course, trying to flirt and make up to them, being robbed of their cigarettes and sweets. The Dutch girls were bewildered by this onslaught and never far from tears and laughter, remembering again and again the hell they had gone through when they were being bombed the night before. Through some kind of misunderstanding, and probably due to my imperfect knowledge of their language, I was suddenly hailed as Prince Bernhard of the Netherlands and drawn into the hospital. I tried to explain, but it seemed to be quite useless in the confusion that followed the discovery. Only after I had been presented to the matron and given all my sweets and most of my cigarettes away, shaken hands with all the pathetic men, women and children who were lying injured in the hospital, was I allowed to resume my ordinary identity.

Our glider pilot flight officers now appeared and began to collect the flight. Burdened with our tremendously heavy rucksacks, we started moving off, cursing and swearing at having to leave our jeeps behind. In single file and directed by a parachute brigade officer we moved towards Arnhem. As the fire was getting heavier now our advance got slower and slower. We guessed that we were the tail of a large column moving down towards Arnhem and were grateful for the long halts, sometimes an hour or two at a time. It was impossible to keep up with the paratroopers who only wore a very small pack. This went on until two o'clock in the morning. We were making less and less progress and heard an increasing volume of fire from the direction of Arnhem. Eventually we got the order to turn round and had to walk back half the way we had advanced that night. We dug in along the railway line, covering the lane along which we had advanced, to be ready for any attack that might come at dawn.

Tuesday

After a few hours the light came up and we went back to a wood near our original RV point. We waited there for orders. At about ten o'clock fighters started passing overhead. We pointed them out to each other and soon the usual argument started. 'They are Spits.' 'Don't be an ass, anyone can see they are Typhoons.' 'Are they hell! Since when have

Typhoons got radial engines?' etc. I kept quiet, as usual when aircraft recce is discussed, completely fascinated how anyone can be so clever as to distinguish one fighter from another at any distance over 1,000ft. Even at that distance I can only tell a Spitfire and a non-Spitfire. I was terribly pleased to be able to join the discussion effectively by pointing out that they were German. This profound observation did not bring me the credit I hoped for as everyone had seen the marking at the same time. Before we had a chance to start a new argument as to what type of German plane it was, their machine guns fired at us. We scattered and ran for shelter. Bullets were hitting ahead of us and until some fools started firing at them with a Bren, they seemed to be uncertain where our position was. The whole string of about fifteen Focke-Wulf 190s and Messerschmitts turned about, losing height at the same time. They had spotted us and for anyone who was not in a slit-trench this was rather unpleasant. They raked us systematically, returning again and again. Once I had found a trench I had not the slightest feeling of fear and realised how little damage this aerial machine-gunning could do. Our casualties were light and not one of our own flight was hit.

We were told that it was our job now to clear the wood and hill of the enemy. Still burdened with our heavy rucksacks and all the equipment and food we had brought with us, we moved up the hill. The firing was becoming more and more intense; we could distinguish heavy and light machine guns and all sorts of explosions. Of course we did not know whether they were ours or the enemy's, because most of us were under fire for the first time. We were told to spread out and go forward to the assault, crawling and hiding behind trees and bushes. At the moment there were only the odd bullets whistling over our heads, but still it was a hard job to keep in line, and the parachute officer had a hell of a time coaxing us forward in anything like a straight formation. Our two officers took the centre and I was the outside man on the left flank. We were about thirty all told.

I tried to work up enthusiasm and hatred for the kind of spirit which I thought was needed when assaulting the enemy. This was no doubt easier for me than for the others. I only had to think back a few years to the time when I was still a German. The past came back to me in flashes, and I had to remind myself that it had really happened, and to me ... I was back in school ... we had racial theory ... the teacher called Hans to the front. He was the funny boy in our class, fat, short-sighted and clumsy, but always jolly, laughing and clowning. The teacher explained that Hans was a Jew and therefore harmful to the pure German stock which he polluted. Everyone laughed and made jokes, but I shall never forget the sight of Hans looking sad for the first time. I suddenly remembered that my father once told me that all my grandparents had been Jews. Should I not join Hans and tell them? But I did not. Perhaps it was that my courage failed me or maybe I did not have the presence of mind. Whenever I thought of it later I felt uncomfortable ... which must have been guilt. Now I realised that this was my unique chance to make up for it to Hans and myself.

Another flash brought me back to the concentration camp. Subconsciously I looked at my hands. They had been bloody and festering from carrying heavy latrine buckets which I was forced to empty with my bare hands. I remembered my 17th birthday at Torgau concentration camp just about ten years ago. I was standing at the window of the dormitory unable to take my eyes off a group of bawling and laughing SA men below having their fun with an elderly Catholic priest by pushing him into the muddy pond where the pigs used to wallow. The priest was up to his shoulders in it, his hair and face were covered in muck so that he looked as if he were wearing a mask with openings only for eyes and mouth. Heavily and slowly he struggled through the thick slimy water towards the edge of the pond, but whenever he reached it the guards pushed him back. They took their time over it so that the whole thing seemed like a nightmare in slow motion. I remembered forcing myself to watch the ghastly spectacle until the priest had finally disappeared in the mud. I felt it was the only thing I could do. If I had not watched I might not have been able to believe it later on.

And I remembered the sickening picture of the smug and conceited SA and SS strutting the streets of my home town, Potsdam, as if they owned the world. I remembered how I used to dream of a most wonderful miracle by which these self-made gods could be deprived of their uniform and power. Then all those who looked up to them would realise that they were the most ignorant and ridiculous bunch of people mankind had ever produced. The realisation that this was the glorious moment in which I could help this dream to come true gave me a feeling of incredible joy and elation. The circle had been completed, justice was being done ... and all this in my time. I was ready to take on anything and anybody that was German.

We reached the top of the hill and the fringe of the wood. In front of us was a large clearing with trees and branches only just felled. This clearing went level for about 200 yards and then dropped steeply and rose again. On the rise, where the brushwood and trees were still growing, were the Germans.

Loud German voices were heard, motor engines were running, and constant machine-gun and rifle fire was directed towards us. I halted a moment before advancing into the clearing to allow the centre and right flank to move up, and when I saw them coming I crawled forward. Bullets were whizzing about us from all directions, but there was no chance of finding out where they came from because the enemy's cover was too good. I found it impossible to advance any further with this damn rucksack on my back so I got rid of it. Looking round, I heard our captain call desperately for our lieutenant and I guessed that the centre of our advance must have had casualties. I was still feeling fighting mad and could not help crawling forward, firing rounds at any movement I could see in front of me. Dodd – one of the pilots of my flight – was on my right only about 10 yards behind me and I shouted to him that I was going forward. Jerry's shouting and yelling was now quite clear; the firing at us seemed to be pretty inaccurate, but consistent. I felt I could not wait for the others. I got up and ran upright down the slope, fell behind cover, then up again and on. I vaguely noticed the intense fire, but all I wanted was to get there, and down I stumbled again between the branches and brushwood of the clearing. I waited and listened for any of our blokes, but all I could hear were terrific bursts of machine-gun fire and the Germans just in front of me. I could understand every word they were saying. Yet my mother tongue seemed like a foreign language to me, for, by the use of a new, overbearing and pompous vocabulary, plus a clipped military way of talking, the Nazis had deprived the German language of much of its sensitiveness and beauty. It sounded terribly ugly and repulsive.

They were quarrelling and swearing at each other. Now they were talking about me and an officer or NCO was telling them to go and search for me because someone had reported my presence. They would not obey, saying that they could not leave the clearing because of the strong enemy fire. Instead, they raked the ground all around me with a machine gun and threw a few hand grenades. I crawled right under the brushwood and saw and heard the bullets splashing the ground and hitting the branches and tree stumps all round me. I was sure this was going to be the end and kicked myself for doing such an idiotic thing; trying to take a strong German position on my own. I swore that if ever I got out of this hopeless position I would never again be such a bloody fool. I lay completely still, bullets whizzing about me. I wondered if I wanted to pray; that is what everybody is supposed to do in a position like this; but I just did not feel like it, and to calm and steady myself I watched a colony of ants go about their well-planned and systematic business.

The Germans were arguing about me again. They did not want to go out to look for me, and so suggested that I must be dead. The more I listened to them, the more I realised what a badly disciplined and poor crowd they were and how easily we could have got them if we had only made a properly planned attack. They were talking now about leaving me until they got their heavier stuff there and I decided that I must at least try to get out of it. I edged back, but every time I moved, either the branches moved with me or my rifle got entangled. I had to do something drastic and quickly, too, for now they knew I was alive. I threw a hand grenade to the left, dropped my rifle and ran for it. For the first few yards nothing much happened and all the firing seemed to go well away from me, but then it came nearer and nearer and I had to flop down. Lying in the brushwood I felt a curious restriction around my legs and discovered that my underpants had somehow slipped down inside my trousers and that I could not possibly make another dash for it. There was nothing else to do but start the complicated business of cutting them in two separate parts. This done, I got up again and made one big dash for the top of the hill and the fringe of the wood from which we originally started our attack. They were still firing at me like mad from the Jerry lines, and now our chaps, who had retired into the wooded area, saw me coming towards them. They also opened up a concentrated fire upon me. Down I went again. I tried to shout, but the moment I moved they let go again and the noise was terrific. I tried again and again and nearly panicked. I just could not make them realise. How I wished that I had a recognition flag, but they were only issued to the first pilots of every crew. I gave up and lay behind the tree, not caring who shot me. It was a ridiculous situation. At last one of the parachute officers realised that something was wrong and stopped firing for a moment. He gave me a chance to call the password. They shouted to me to get up and raise my hands. Up I got. They went on firing, so I had to lie down again. It was only then I realised that I was still clutching a hand grenade; since I had abandoned my rifle this was my only weapon; I had forgotten all about it when I put up my hands. They shouted at me to put it down and fired again. Only after telling them exactly who I was, was I allowed to go forward.

I told the parachute officer about the bad morale of the Germans and that if he attacked now he would be sure to bowl them over. I said the same to the men who were near me and was asked to come with them into the new assault. I did not like this at all because I thought that it was tempting providence too far, but after encouraging them in this way I could hardly stay behind. So off I went again, to the fringe of the wood to take up an assault position. I vaguely heard engine noises from the German side, then a terrific crash and black smoke and sand flew up in the air; the Germans were using tanks or SP guns. The second shot went straight into our line of attack and six men on the left flank were knocked out. There was nothing to do but retire to our original position, and I linked up with our flight, who were digging in on the rear slopes of the hill.

They told me that they had four casualties when we started our first attack, and that our lieutenant was still out there in the brushwood seriously injured. I wanted to go back and look for him as I had seen something moving just when we were going in for the second attack, and I was certain that I could find him if he was still alive. Two of the blokes came with me and waited on the fringe of the wood while I searched the brushwood. I found him at last, but he was dead and it was not worthwhile getting the other chaps along to pull him out. We could definitely have had casualties doing so. Back to our old positions and digging in, we started making some tea and things as we had not had anything except biscuits, chocolate and sweets since Monday morning. Just before four o'clock the first bombers appeared on the horizon. They came slowly towards us in a seemingly never-ending stream. There were Stirlings, Halifaxes and Dakotas, many of them with gliders in tow. The whole sky above us was filled, like a moving ceiling, just below the cloud base. It was an awe-inspiring show of might; it seemed impossible that anything could deter this steadily advancing flow from its predetermined route. The deep all-filling drone of their hundreds and hundreds of engines made you feel the spell even more. Then, as if by a single word of command, scores of ack-ack batteries opened up. The throb of the engines was suddenly swamped by the furious bark of the guns. The ear-splitting fury of the attack from the ground was indescribable. The stately procession of bombers carried on without seeming to take any notice for a few seconds; then these giants began lumbering out of the way, diving, banking, climbing. It seemed so undignified and pathetically clumsy, somehow. They were so helpless; I have never seen anything to illustrate the word 'helpless' more horribly. Now the sky was chaos: puffs of exploding shells, bombers alight, bombers plunging towards the earth, gliders casting off and banking steeply, and in between all this an irregular thick pattern of parachutes; men and supplies floating down. We of the first and second lifts thanked God that we were already on the ground.

But we never finished that meal, as the order came to move back to Wolfheze. Five minutes later we joined a stream of troops moving back slowly; we realised that this was a retreat. We were evacuating the woods and hills we had been about to dig into. It was a long stream of troops, of all units, walking rather quietly down the slope. Disorganisation started when we had to cross an open field which led to the railway lines. This field was under rather inaccurate German fire, but still it made everybody run. When we reached the other side we were not an organised body of men. The men had lost their officers and the officers their men. But everyone was disciplined and quiet in themselves, and there was no shouting or pushing. We helped each other along, said 'sorry'; we were just dazed and found the retreat rather incomprehensible.

We got to Wolfheze cross-roads where troops were lining the road and bunching together everywhere. It was impossible to find out what the plan was and what we were going to do. I felt terribly uncomfortable as I knew the Germans were so near, and, disorganised as we were, we could not possibly put up any resistance. There were twenty of us, mostly from our own flight, and as we could get no clear-cut information of what to do, I thought it best to retreat towards the river, where we hoped the Second Army would soon relieve us. I knew that it might be wrong to go on retreating off our own bat, but my feeling of uneasiness at remaining bunched together near the railway lines and cross-roads was so strong that I could not do anything else.

We started using our compass and maps, and took a road which was supposed to lead to the river just right of Arnhem. It became very quiet the farther we got. We could hear distant firing, both of small arms and artillery from either side of the road, but we did not meet a soul. Most of the chaps felt that this was too uncertain and the silence depressed them. They said they would return to Wolfheze and I didn't argue as I did not want to be responsible for them. This move of mine was purely instinctive and there was no plan or reason attached to it. Only Dodd stayed with me, and we carried on along the lonely road. I never saw any of the people I left in Wolfheze again. They did not come away with us at the final pulling out of the First Airborne Division, so I do not know what happened to them, nor could any of the people I asked tell me. We walked on just inside the woods, never losing sight of the main road.

Just as it was getting dusk a string of jeeps came racing along. We hailed them – it was a reconnaissance patrol and they were glad to take us as they had had some casualties and their numbers needed making up. We raced through the woods at 60 miles an hour, sitting sideways on the jeeps and covering the woods on either side with our guns. We were lucky and got right through to the recce HQ without meeting any of the enemy. It was getting dark now and they were already dug in for the night, so all we had to do was to get some tins of food and blankets. Of course we had nothing at all, having had to leave our rucksacks behind, but the recce blokes gave us plenty of everything, and we did not have to take turns on guard. I got myself a Sten gun, plenty of filled mags and had a whole night's sleep in my slit-trench.

Wednesday

After breakfast they called for a patrol to push forward on to Arnhem railway bridge. Dodd and myself were asked to go. Off we raced again, and after ten minutes the first jeep encountered fire and pulled into the side. From here our advance was very slow. Everybody, except the driver, got out and worked their way forward on both sides of the road. On one side was a thick wood and on the other houses with gardens in front of them.

Progress was slow, now that the jeeps were parked in the gardens along the side streets, and two officers and twelve of us moved into the wood on the right. The others advanced through the gardens on the left. We heard German voices shouting and bawling just on the right of us, and also saw some of them moving backwards towards the bridge. We exchanged fire, but it was pretty inaccurate on both sides and we kept on, moving forward quite steadily. Then I heard the same engine noise as yesterday, and not long after the same old crash and thud. The tank was moving forward and firing shells into the houses across the road on our left. The chaps across the road stopped advancing and we manned a defensive position in the wood. We put a Piat gun in position and lay there waiting. The noise of the tank got nearer and nearer, and so did the shells hitting the houses opposite us.

A tank advancing firing shells is the most frightening thing imaginable, and of all the experiences I had later on I was never more frightened than at this moment. I believe that this is what makes a tank such a formidable weapon. We only had a little Piat gun just 3ft long. The feeling of helplessness and fear became stronger and stronger the nearer the tank came. And at the same time the German infantry was working round us, obviously screening the tank. The voices and shouts seemed to be all about us. Just in front of us someone had thrown a smoke grenade on the road and, before we knew what had happened, the recce had crossed the road. We saw them start up their jeeps and off they went. There was only the Piat gunner, his Number Two, Dodd and myself left. We decided to try and make our way back to the recce HQ.

We could not cross the road any more as the tanks were firing down it like mad, but we knew that we had to cross it if we wanted to reach our destination. We withdrew through the wood parallel with the road, firing our guns to keep the Germans away, and throwing a hand grenade now and then to frighten them more than anything. Like that, we got as far as the cross-roads which were about 4 miles from our HQ. Here there was no chance of crossing the road or of getting any farther, as we heard German voices from all directions.

There was nothing left but to hide and hope for the best. We were now in the back gardens of some houses near the cross-roads. The two recce blokes hid in a thicket just behind, while Dodd and I approached a house. We looked into a little shed and thought of going into the house and hiding in the cellar, but we did not know whether the civilians were Dutch Nazis or friendly. Just then we heard footsteps approaching. I spotted the rubbish dump belonging to the house – it was a little pit, 4ft by 2ft and about 3ft deep, neatly hidden by some shrubs just about 10ft from the door. We crouched in it, and had just got our heads below the shrubs when the Germans came in to search the garden. They looked into the shed and through the house and, after satisfying themselves that none of us were hiding there, they just stood about talking in very loud voices and giving each other orders which no one obeyed. Most of the time they stood 2 or 3ft from us. It was terrible in this dug-out. The decaying garbage stank and gradually seeped through our trousers. Our limbs got cramped and we felt pins and needles everywhere. There was no chance of moving at all, but the desire to move was irresistible. Then I discovered that my arms reached the sides of the trench and, by moving my hands, I could undercut the trench and so clear a little space for our feet. Dodd followed my example, and after an hour's scratching, things became a little more bearable. After what seemed like ten years, the Jerries left the garden. We were just contemplating making a move when we saw a glider pilot jump out of the back door of one house and into the next. Oh, the agony when we tried to get out of our pit! But the joy of seeing a friend made us jump out of the pit and into the house.

The glider pilot told us that we must get away quickly. He had hidden in the attic of the house next door and from there had shot the whole crew of a tank that was standing on the corner of the cross-roads. He had just got out of the house when they came to search it. Obviously they would search this one, too. All three of us scrambled back through the garden to find the thicket where the two recce blokes had hidden. They were still there and another six men besides. We crawled in and lay down exhausted, to wait for the night. It was only just afternoon, and it looked like being a pretty long wait. Very soon mortars and shells started whistling over us and the firing became more and more intense. We lay there, completely silent, on our stomachs, as near to the ground as possible. There was no chance of digging in or of taking shelter in this thicket surrounded by Germans. Lying there so inactive made us all desperately frightened. But we were lucky, for after about two hours the bombardment ceased, and we started whispering and making plans. A patrol of ten men approached us; their quiet and disciplined movements betrayed them straight away as British. It was Lieutenant W with ten glider pilots from different squadrons. He asked us all if we wished to join his little group and try to make our way back to the glider pilot HQ. We three were only too pleased. The recce blokes preferred to stay in the woods until nightfall.

Fortunately Mr W knew the geography of this place quite well and, following his compass, led us through the woods. Again, long before we met the Germans, we were warned of their presence by their shouting and calling to each other. We took up a defensive formation, but carried on in the same direction. Then we came upon them about 100 yards in front of us. They were filing into an isolated house, surrounded by a wall, which was standing amidst the trees. We let go and several of them fell down injured. The others started rushing out from the house, colliding with the ones trying to get in. They were apparently helpless, and our two Bren guns were just getting into position to let go into their midst, when I asked Mr W if he and the others could cover me. I got up and walked straight towards the Jerries, clutching my Sten gun and feeling amazingly safe and powerful. I shouted '*Hande hoch!*' (Hands up!) and told them that the Second Army was just coming up, that they were hopelessly surrounded, and would they come and give themselves up. Very slowly they started filing out of the house into the road and I walked nearer; suddenly an officer appeared and furiously ordered them back.

We started firing and hit some of them. But they started to return our fire from the windows of the house, and of course, as there must have been about fifty of them inside, we could not hope to beat them now that the officer was organising the defence. We decided to withdraw. Unfortunately the young glider pilot, who had just before killed the whole crew of a tank single-handedly, got shot in the stomach and Mr W detailed two chaps to take him back to our Divisional HQ while we kept Jerry busy in the house.

Eventually we reached a place on the outskirts of Arnhem called Oosterbeek. This was mostly occupied by our troops, and here were the Division and Brigade HQs. It was to be our defence perimeter during the rest of the operation. It was glorious to see British troops again, and I went into the first house to refill my empty Sten mags. They were making tea and of course that was heaven, and I had to remain and have a mess tin filled with some biscuits and jam. After I had loaded myself with food and ammunition I found I had lost Mr W. I asked my way to the Division HQ and from there got to the Glider Pilot Regiment.

There were quite a few of them dug in in a large park. They were guarding the Division HQ, which was a very massive hotel [the Hartenstein] surrounded by large outbuildings, hothouses, etc. I think the place had been very famous. Until a few days previously it had harboured the German HQ for this area. There were not many people I knew in the trenches, as they came from different squadrons, but as there was no sign of anyone from my own flight, I decided to remain with them. I was much too tired to go round looking, and anyhow what did it matter where one fought? They were all living in extremely deep slit-trenches with roofs and branches and all sorts of ingenious contraptions covering them. I heard that I had arrived during the third lull that they had had. The mortar fire was increasing all the time, and I had to find myself a slit-trench for the night. Eventually I did find six blokes of my flight, but none of them had heard anything of the rest of the flight or of our officers. I joined them, and one of the pilots offered to share his slit-trench with me as it was too late to dig one of my own. I also met Vic Wade, who was my best friend back at the station, and several others. I felt that I could not possibly spend the night in the trench with Jimmy Plant as there was not room for two, so I moved a bed out of the gardener's house, which was just behind us and very badly knocked about. I put it right next to the trench and slept comfortably in sheets on a feather bed.

For some reason or another that night the mortaring was rather light and mainly on the far side of the perimeter. Or maybe I slept so deeply that I was not aware of the danger. But not once during the night did I have to roll out of bed into the trench.

Thursday

After the dawn stand-to, we were just getting breakfast ready, which meant heating the prepared tins of bacon, beans, steak and kidney pie or anything which we could scrounge, when the most intense mortaring of the Division HQ started. It was our area which took the brunt of this attack and everyone crouched in the trenches. It was quite terrifying. Any kind of danger is a hundred times more frightening when one has to remain inactive, and all we could do was to crouch in our trenches, feeling the mortar fire creep nearer and nearer. Someone who could not reach his own trench in time jumped in on top of Jimmy Plant and myself but we did not dare to move as the hits were falling right amongst us. There was a thunderous crash ... the walls of the trench crumbled and we were covered with branches and sand – I felt dazed and was certain that my eardrums must have burst. We kept still for several minutes, just to make sure that we were all in one piece and asked each other if we were all right. Then we clambered out and found that nothing had happened to us. The crater was just an arm's length from our trench and I could touch the centre from where I stood. The mortar had gone right through my beautiful bed and shot it back through the window into the house. There was no more breakfast for any of us – not that we would have been able to eat any. The barrage passed on towards the Division HQ building. An officer asked for volunteers to go on a patrol. I was glad to get out of this, and off we went through the park and the perimeter.

The main road passed along one side of our perimeter and was crossed by another road which formed the other border of our position. This smaller road had a row of houses along its left side, which led away from the perimeter; on the right side of the road was a wood with a few houses standing in the trees. At the back of the houses on the left-hand side was a vegetable plantation 200 yards wide, belonging jointly to this row of houses and the row opposite, making a large square which was surrounded on three sides by the backs of the houses. There were about twenty small plots and gardens with fruit trees, runner beans and all sorts of vegetables. We advanced through these gardens to find out if the houses on both sides were occupied. This was a very slow and laborious undertaking. We crawled towards a house, waited and listened, crawled again, waited, and so on and so on. As we didn't seem to make any headway, I began to get impatient. I got up, went to the next house, knocked on the door, went in and shouted *'Jemand hier?'* (Anyone here?) and as there wasn't, we went through the rooms to make sure. This went on admirably from house to house, and sped up operations considerably. In one house, some civilians answered my call and led me to a cellar where there was a wounded British paratrooper. We sent someone back to fetch a medical orderly. We made sure of all the buildings along the cabbage patch and crossed over the road at the far side of the plantation, which ran parallel to the main road, where my polite enquiry resulted in a stampede – not towards us, but away from us, Jerry, apparently did not care for visitors, and made his exit, leaving a machine gun and various other equipment. There seemed to be about six of them, who bolted into the house opposite. We lobbed hand grenades after them before they could take cover, and one of them was laid out. I had run upstairs and could see them inside a trench and behind hedges, trying to crawl away. I fired, and hit another, then went on lobbing hand grenades from my position overlooking their trench. Captain Z sent me along the road to try to link up with a party which the colonel had led, to report Jerry's position, it was a pity we couldn't drive them out of more of the houses, but we were not supposed to get into a serious fight. We had to await orders.

So we remained in one house and were told later to retire back into the vegetable plantation and occupy the houses on the right. The row on the other side was to be occupied by the KOSB (King's Own Scottish Borderers). We split our own party into two and took over the corner house at the far end of the street and the corner house on the large road nearest to our defence perimeter. We stayed in this street until the whole Division withdrew on the Monday. Our position was in the lower corner house. The commanding officer of our party was a big young man with a fair, stately moustache [this was Captain Ogilvie]. He wore a kilt. With him, there were five other officers and about fifty other glider pilots. The end house was occupied by two officers and about ten to fifteen men. The moment we moved in, Jerry started to harass us. Apparently he had had the same idea, just a bit too late. He was now sniping at us from the woods and the few houses which faced the road opposite our row. We could not walk from one house to another, or appear in any of the front rooms, without someone having a pot at us.

We started to settle down in our two houses; digging trenches in the front garden and setting up Bren-gun positions which would cover as much of the road as possible, but we were too many in each house and much too disorganised, and there were only a few hours of daylight left. No one did anything about barricading the front windows and occupying the twelve houses and gardens which lay between us. I felt that something must be done very quickly if we were going to survive that night, for Jerry was becoming more and more active in the wood only 20 yards away. Everyone was clustering together in the two houses and nothing was being done. I went to see Captain Z and, with the help of a lieutenant and others, we worked out a plan to occupy at least every second house until we got enough reinforcements to defend every house in the street. Many of the things I had read in descriptions of street fighting in Spain came back to me and they proved invaluable later on. We barricaded the front windows so that Jerry could not throw hand grenades into them. We dug communication trenches from one house to another and chiselled holes in brick walls. We put up branches for cover and did everything possible to have safe, invisible communications along the whole street.

Our main idea was that, if any house could not hold out, the occupants could fall back upon the next house and reinforce it; besides, it was essential to know what was going on in every part of our street.

That night I stayed in the lower corner house, firing magazine after magazine up the road with our Bren gun to prevent Jerry crossing over from the wood. They fired back fiercely all night but did not dare to attack us. As soon as the first light came we went on with our preparations, and also got some reinforcements from the Division HQ. In addition, some of the Polish Parachute Brigade, which had landed on the other side of the river and fought its way over, were detailed to our street. They took up positions in three houses in the middle of the row. We glider pilots could not concentrate in the top six houses. It was a relief after having the responsibility of the whole street on our shoulders, especially as we had had several casualties during the night. There was not a day when a few of us weren't knocked out. Our Glider

Pilot HQ was in the third house from the top, and there were ten of us holding it. The street was not completely occupied.

Friday

During the morning the first German SP gun started moving around the top cross-road. We heard the engine revving and the Jerries shouting before the attack started.

The immediate job was to put our Piat gun in a position where it could dominate the road and prevent the SP gun from moving down. We found an ideal place which Jerry never spotted the whole time we were there. We decided to fire the bomb through a little hole in the roof of the attic. The noise of the tank got louder. We could hear the tracks squeaking and grinding along the road. Then the first shots were fired and tore away some bricks from the front of the houses. From the wood opposite came the splutter of a heavy Spandau machine gun and a hail of light machine-gun and rifle bullets came across the road. Jerry was also using small armour-piercing shells, which could penetrate clean through a house. This was all preparation for the SP. Until we could see the tank coming, we used the Bren gun from our high lookout and sprayed the cross-roads and the wood opposite us with continuous fire. Germans were moving about, unaware that they could be seen, and many of our bullets found their mark. Lieutenant X came up to the attic and offered to fire the Piat gun. He had just a little more experience with it than I had, as he had once fired a practice shot while I had once been shown how to load it. A wave of disgust rose in me against those petty-minded officers and sergeant-majors who had wasted weeks, even months, with drill and kit inspection, making us lay out our boots, toothbrushes, knives, forks and other kit, as per regulation. Why couldn't they have taught us about house-to-house fighting and the Piat gun? But then, drill, and lining up of beds and blankets, occupies the greatest number of men with the least effort.

Lieutenant X waited until we could see the tank clearly – it could not have been more than 100 yards from us. Then he fired the first round. It was the greatest joy I had felt for a long time when we heard and saw the terrific explosion this little weapon produced, a relief which could hardly be described.

To sit there waiting for the monster needed all our patience and strength for we had no idea what this little apparatus could do against it. Lieutenant X was covered with dust and thrown against the other wall by the recoil as the bomb left the Piat. I had taken position next to him ready to jump forward and look through the hole to see where it hit. The direction was perfect, but it fell about 20 yards short. The SP stopped immediately and, by the time Lieutenant X had shaken himself and got back into position, I had reloaded. We fired another four or five shots, and the Jerries obviously couldn't decide whether it was one of our very few anti-tank guns or what, nor where it came from. They were firing straight ahead and at our side of the street, but all the shells went well past us, hitting houses and trees farther down. Apparently they hadn't the faintest idea that we were only 100 yards away, sitting with our popgun in the attic of the nearest house.

The SP retired about 50 yards, far enough to be out of our range. Now that we knew the value of our Piat we took it down into our safest back room, together with the bombs, and I continued firing the Bren gun through the hole in the roof to cover the cross-roads and make it hot for any Jerry who tried to cross into the woods opposite.

The firing and sniping went on. Suddenly it got quiet and, from our three hospital buildings on the lower cross-roads just outside our perimeter, appeared two of our jeeps with large Red Cross flags. Whatever I personally felt about the Germans, I must give them their due; in this Arnhem action they couldn't have kept more strictly to the Geneva Convention, and this was confirmed by everyone I talked to. In a way their behaviour was so deliberate and precise that there must have been a policy behind it. Not once did I hear of any Red Cross men or jeeps being deliberately fired on, even when they appeared on the most contested road in the midst of the heaviest fighting. There were several men

I spoke to who were taken prisoner by fast-advancing tanks and whom Jerry allowed to go back to our lines with a kind of slap on the back. The work of the Red Cross personnel was wonderful. Tremendous courage and self-sacrifice is necessary to drive or walk out of hospital gates and along a road which is under fire from excited and heated troops. I was told that the casualties of the Red Cross were at least as heavy as those of the fighting men. It is difficult to see from 300 yards ahead if a man has a Red Cross armlet. Very often they couldn't be seen through our camouflage. Shooting at them by the Germans, or even by our own men, could not possibly be avoided.

The casualties in the lower houses of our street must have been very heavy as the jeeps soon reappeared with their stretchers occupied and wounded with first-aid bandages sitting on the sides. Other wounded were walking behind a Red Cross flag carried by one of the orderlies.

The firing started up the moment the Red Cross party disappeared into our lines and stopped the moment they came out again.

I left my post for a minute to get some rest. I wanted to see how our house had stood up to the attack and whether there had been any casualties. The situation seemed to be well in hand and we were still in possession of the corner house, though the tank had been more or less level with it when we had stopped it. The occupants told us of the relief they had felt when they saw the first Piat bomb hit the road. They thought we had damaged one of the tracks, as the SP retired in jerks as if it was out of control; it might even have been pulled back by a recovery tractor. Everyone reported that the sniping and firing was very bad, and we decided that our communications between the houses must be improved. We needed deeper and longer trenches, camouflaged with branches, to allow us to dig in comparative safety.

Three of us went round systematically barricading our front windows and doors. We saw the effect of hand grenades in the front rooms. Barricading was an uncomfortable job; we had to use the beautiful antique furniture, which must have been of great value, to block up the windows and doors. All this stuff was going to be wrecked at the next attack. The front rooms and facades of the houses suffered each time.

This job of barricading had to be redone every time we had a respite, though we never had a real respite as rifle and machine-gun sniping kept on consistently and perseveringly.

One got skilled in avoiding being hit, and as time went on our casualties became fewer, though we were desperately tired and thought less about personal danger. But we had acquired a kind of sixth sense and somehow did the right things automatically. In moments of half-dozing, while manning my attic position, I felt terribly pleased and grateful for this newly discovered ability. No one can know beforehand or can influence their reactions to great personal danger. And this feeling of pride and pleasure compensated a little for the hatefulness of the whole bloody business. I hate war. I can't stop thinking of the friends and relatives of anyone who has been hit. I know the Germans. I have seen them do the most vile and frightful things. I know that they have destroyed millions of Jews and political opponents. But I do not enjoy killing or wounding anyone. Once I'm forced to fight, however, the whole affair becomes a matter of skill and a job that needs all my powers of concentration. I no longer consider the effect it has on my opponent.

We spent the rest of the morning watching carefully, trying to keep the snipers quiet and improving our position. Soon we began to feel very hungry and the food problem had to be seriously considered. I joined Graham in search of food. We hadn't had a single issue since we had landed. Most of us had lost our rucksacks, and those who had managed to save theirs had shared all their supplies. We were very lucky to find large stores of tinned food and provisions in the houses. I didn't know how the fellows managed who had to fight in the wooded country and fields, but we in the houses did very well. With a little bit of searching, we got the most magnificent meals together; these would have gone down well even at home. Every Dutch house has a store of food preserves in all sorts of glass and earthenware containers. Trust the Dutch! In other houses they were all helping themselves liberally to bottled tomatoes, French beans and other vegetables, avoiding the preserves which had an unappetising look. I went round all the cellars, asking if I could take a few preserves with me. Our expandable battle smock could harbour at least three or four of the 'unappetising' preserves, and in my hand I carried a respectable jar. The first time I came back with my spoil there was a general outcry and I was told I would have to cook and eat it myself. I had the utmost difficulty in persuading some of the chaps to eat what they called 'Continental concoctions' but I finally did persuade them that, although they looked like bottled medical specimens, they were really very good to eat. As indeed they were.

Very soon everyone was eating like mad and had completely forgotten the war, the Second Army and the Jerry tanks. They were stuffing themselves with fried chicken, tenderloin of pork and beefsteak. We laid a fire with bricks in the middle of the kitchen floor. By this time everyone was enthusiastic about the food and was 'digging for victory' at the risk of their lives, in the plantation. This resulted in a flare-up of German fire as they thought we were going to launch an attack. All our rich harvest was thrown into a large slop pail and cooked on the open fire. It was delicious.

We felt like a rest, but the now familiar though still ghastly sound of engines and tracks was heard again. Up dashed Lieutenant X to our beloved Piat gun and, loaded with bombs, I followed him to our attic position. With the aid of binoculars, we could just see a mass of branches and trees with movement behind it. We let go with a spray of Bren bullets to make it hot for any troops who might be hiding behind the tank or whatever was at the back of the approaching foliage. Now we knew that a lot of bullets and noise whizzing round Jerry, even if they were very inaccurate, would help us, and would discourage him from any personal assault or attack. He was desperately afraid of us, and that was one of the reasons why we held out until our planned withdrawal; that and our faith in the Second Army. There were so many proofs of this fear that there could be no doubt about it. First, when I was lying in front of the German lines and they did not dare advance or attack until their tank came. Again, when against ten of us nearly fifty of them in a strong position almost gave themselves up and were only stopped by an officer. Then, their constant shouting and bawling at night, for no other reason than to give themselves confidence. Also I had an opportunity to interpret for a parachute major, who did the preliminary interrogation of two German prisoners as soon as they were marched in.

They belonged to the SS Panzer Grenadiers and gave us their regiment and number of unit, etc. They said they had been in the army for six weeks and this was their first action. They were both about 40, and obviously had no intention of fighting anyone or anything. They said that they knew the war was lost for Germany, and when I asked them why they fought us, when it was useless, they said that they had no choice and would have been shot by their own people at the slightest sign of refusal. I heard this sort of report over and over again.

If there had not been a sprinkling of first-class and fanatical officers and NCOs in this division, no fight would have been possible. But even with the present state of affairs, it was ridiculous that they did not wipe us out within a few hours. This Panzer division, with tanks, mobile guns, flamethrowers, very close Focke-Wulf support and the heaviest and most concentrated ack-ack seen by any of the RAF pilots whom I met later on at the 'drome, and even mobile loudspeakers with trained German propagandists spouting in English, never dared to change over to direct assault or succeeded in penetrating our perimeter. No body of men, with only small arms as we had, could possibly have withstood a German Panzer of the old material.

Slowly the mass of foliage drew nearer and started to fire down the street at the lower houses. We could now see the immense tracks and from their size we thought that this was probably a SP gun. We couldn't do anything for quite a long time, and even decided that it would be inadvisable to go on firing the Bren, as this might give our Piat position away. Only when the SP had got to within about 100 yards of us could we be active, and that would be a God-sent relief. Waiting and watching the gun approach was almost unbearable. It made a terrific noise and smoke each time it fired, and we could hear the clatter of glass and masonry whenever one of the shells hit a façade down the street.

We had decided that we must try and increase the range of our bomb by elevating the Piat still further, using it more or less like a mortar. Just in front of our attic were the branches of a large tree, and about a couple of yards away was the corner of the next house. Our first attempt proved pretty inaccurate, and the bomb must have hit one of the tree branches and been diverted on to the corner of the house next door where it exploded with a terrific blast. This really shook us badly, though mainly Lieutenant X who was standing right behind the Piat; he was flung against the back wall of the attic, and I saw him covered with dust and looking very pale, crawling towards the stairs. There was nothing seriously wrong with him, thank God, it was just the effect of the blast that had winded and concussed him; for the moment he didn't seem to know what had happened. He went off down the stairs, and I continued firing bomb after bomb the old well-tried way. I yelled for someone downstairs to bring up more bombs and insert the fuses.

Well, we had done it again ... I don't know how much we had damaged the SP, but it stopped firing and withdrew out of range, slowly and critically. It stopped just behind the rise in the street, and I could still see it from the top of our house, but the Jerries must have thought that any kind of anti-tank gun firing from the bottom corner couldn't see it. Men started busying themselves round it, until all the top houses opened up with their Stens, Brens and rifles and drove them out of sight.

As usual, after any kind of concentrated attack, the firing suddenly ceased when the Jerries, and we alike, brought out our ambulances to collect the wounded. Now that it was quiet, the old background of rumbling artillery was audible. We never knew whether this was ours or the Germans', but liked to think it was the Second Army shelling the German rear. There were always the rapid mortar explosions, falling mainly on the perimeter and Division HQ, and bursts of machine-gun and small-arms fire, but these noises were so continuous through all the days and nights that we held this street that after a while we didn't hear anything, and everything seemed perfectly normal.

We were always busy at our posts, in the trenches in front of the houses, in the attics, at the windows, and in little firing positions made by removing a brick from the outside wall. We covered all the approaches in the gardens, the street or the woods. Reliefs were fixed mostly by mutual arrangement. Food was collected and cooked in the lull between the heavy arms attacks and attempted assaults. We were now busy again patching up all the breaches in the barricades round our house. They had to be complete for the night, and everyone felt much more confident about holding out than we had done the previous night.

By now Lieutenant X had recovered from his shock, and we visited other houses, making arrangements. He suggested an hour or two's nap before the inevitable intensification of activity on both sides at dusk. We went into the only still completely furnished room, called 'The Officers' Room' because in it, Captain Z, the commanding officer, who had been wounded in the arm, was usually lying in a luxuriously heavy Continental bed, with sheets and pillowcase, covered with a quilted eiderdown. He lay fully dressed in kilt, sporran, boots and beret. Full of confidence and optimism, he received everyone on his bed, and it was difficult to know how far he realised the danger of our position. In addition to Captain Z there were usually a few other officers – men from the other houses, making plans and decisions in a nonchalant way.

Only once during the whole action did I hear any one of the officers really lose his temper, and that was when trying to wake up the sergeant-major, who had his HQ under the best table. It took me several visits to the Officers' Room to discover him, since the huge lace tablecloth, which overhung a purple velvet cover, hid him rather effectively. And that was exactly what the sergeant-major wanted. His aggressive nature on the parade ground and in the barracks had changed entirely and he did not want people to notice. Everybody did, of course, know about it, but nobody could do anything, and in any case, there were plenty of us who could do his job.

Nobody could escape the spell of this room. The shutters were closed, and the candlelight, bed, couch and easy chairs gave it such a homely atmosphere. It needed great willpower to leave this room and carry on with any job. But at least it made you feel that somewhere there was still peace and homeliness. Lieutenant X and I had our nap on some cushions on the floor in a position that was safe against sniping or blast.

It was now dusk and everyone was mobilised for the stand-to. The supper was simmering in the pail and the chaps came down, as each was relieved to have a plate of hot stew and some delicious preserved fruit, with sweet condensed milk – the winter rations of the Dutch people. I was having my supper when Lieutenant X came into the kitchen and asked me to come to the Officers' Room. I took my plate and went in. Captain Z had just seen the brigadier and our own colonel [Brigadier Hackett and Colonel Murray], and had been asked to send patrols out to discover the assembly point of the German armour, which was harassing our section, and to find out which of the houses at the end of our street opposite the top corner house were occupied by Germans. They asked me if I would like the job and how I would tackle it. I thought it best to take as few people with me as possible as the important thing was to get near Jerry undetected, one chap to be with me the whole time, and a small group waiting on the other fringe of the plantation where I would cross into Jerryland, to give me covering fire in case I should run into trouble and have to withdraw quickly.

I now had to find someone to come with me and I thought immediately of Sergeant Graham. He had taken a lot of pains and responsibilities far above his rank and seemed more or less to have the same ideas about the situation as I had. I brought him into the Officers' Room and we discussed our plan. We were going out on three patrols at ten, twelve and two o'clock, on three different points of the plantation. Next we had to go searching for civvy shoes or running shoes, and clean our Sten guns and ammo. We slept until half-past nine, when we were supposed to meet our covering party, which Mr T [an officer] had offered to supply from his corner house. We never met his party. I don't know what happened or where it waited for us, but we were much too impatient to go searching for it and stumbling about, giving away our position and intentions. We crossed the plantation over the street to the next block of houses – slowly tiptoeing in the shadow of the houses. Sergeant Graham stayed about 20 yards behind me, following up every time I stopped for any length of time. German voices became audible now and we could hear engines running. They seemed to be manoeuvring their transport and armour. Then we heard Germans walking through gardens. We lay just where we were. We began to distinguish a general movement from right to left. The German transport was moving to a point about a mile away, screened by small detachments moving parallel to them. We tried to get nearer the machines, but unfortunately we ran into a body of the enemy who immediately opened up on us, even though I'm sure they couldn't have seen us. We decided that we had better withdraw this time and try to get farther on our next patrol. We now had some idea of the lie of the land up to where we had to withdraw; that would save us a lot of time at twelve o'clock. Besides, we were supposed to cross into our own lines at half-past ten as our men had been given that time for our return and told to hold their fire. Now for a sleep, and by twelve o'clock we were ready once more. The troops covering the lines we had to cross were warned. We had realised that we must have longer for this patrol, so our time limit to get back was extended from 12 to 1.30.

We crossed over as quietly as possible and worked our way through the maze of the back gardens, among outhouses, shrubs and orchards. Soon we had passed far beyond the place where we had met the German patrol before; the talking was more distant now. We made slow and very careful progress. It was pretty nerve-racking, worming our way along, silently stopping every few minutes to listen for German footsteps and noises, which were to be heard now at much longer intervals. We could hear a more or less continuous noise of spades digging, and this was the direction we took. The noise led us to a thick clipped hedge, and we tried to wriggle through, but we got stuck again and again by our Sten guns and things catching in the dense undergrowth. Then we found a square tunnel cut right underneath the hedge, through which we crawled. We emerged into a large open space, bounded on one side by the hedge and on the other by what looked like the outbuildings of a big country house. In the middle of the space were two immense oak trees with circular benches round their boles. A path ran up for about 100ft, and at the top we saw the silhouettes of two Germans digging and whispering monotonously. They looked quite unreal, as if they were standing on the walls of some fairy castle. This hill was not marked on any of our maps, and we had had no knowledge of its existence.

Here, obviously, was the German strong-point. From it they sent out troops and vehicles to dominate our position from the woods and houses opposite. We now realised that they occupied these houses and the wood during the daytime and withdrew their men and vehicles into this strong-point at night. At the moment it looked much more like a fairy castle. Graham thought it would be a good idea to kill the two Jerries up there. They were less than 100 yards away. But our withdrawal through the gardens and orchards was going to be difficult in any case, and from their dominating position they would be able to plaster us with hand grenades, so I thought it not worth the risk as our information might be of great importance.

When we got back we were surprised to find it was well after two. We were glad to think that we now had a good excuse not to go on the third patrol. We got through our own lines without being accosted; either everyone was fast asleep, or perhaps they were waiting for us.

In our cosy Officers' Room we were surprised and rather touched to find that they hadn't all completely passed out, but were really quite anxious about us and our long stay.

Under normal conditions our report would have been of immense value, for now we could pinpoint the German position accurately on a map for artillery and mortar fire. But then we realised that, as we had no mortars or artillery, there was nothing we could do about it, although it was interesting from a military point of view. It was valuable, however, to know that Jerry left his houses every night and retired to safe, prepared positions. Obviously the men refused to spend the night so near the British lines. That was a compliment. If only we'd had the men, we could have given the Germans a lovely surprise by infiltrating into their houses and establishing ourselves there by dawn. Still, Captain Z must have been pleased, and attached quite a bit of importance to our report, for he condescended to sit up in bed, twirl his handsome moustache, and say, 'Well done, chaps. Good show. You've got to see the Brigadier tomorrow ...' Graham and I gathered up some cushions and blankets from the floor, found a corner and passed out until stand-to at dawn.

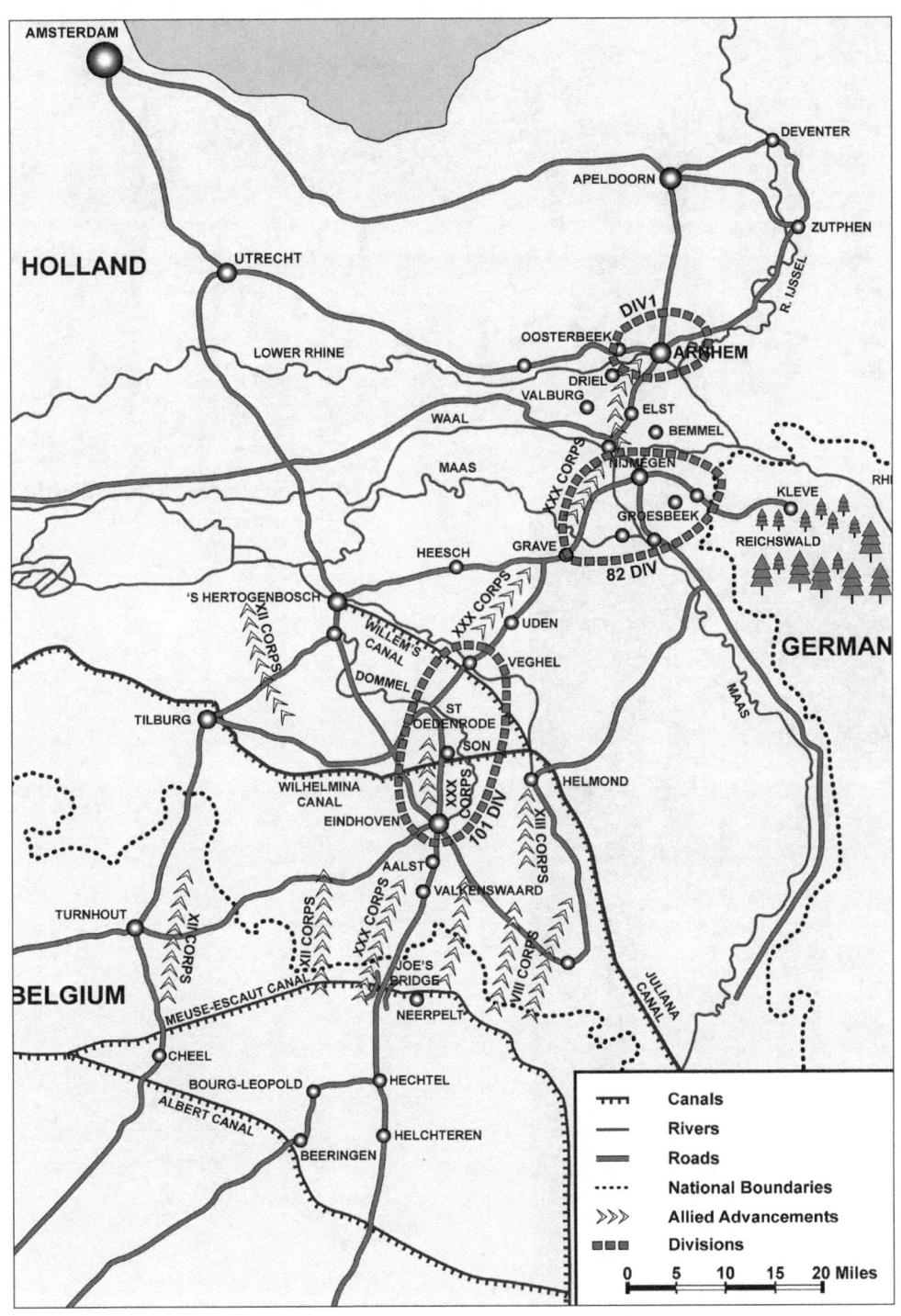

The plan for Operation 'Market Garden', September 1944.

Louis (second from left) in the Pioneer Corps, 1940.

Louis in flying kit, 1944.

Winrich Behr at La Roche Guyon, spring 1944.

Winrich Behr with Field Marshal Rommel, Normandy, February 1944.

Sergeant Louis Hagen (left), Glider Pilot Regiment, with his two brothers, 1944.

The air armada for 'Market Garden' deployed the two American and one British airborne divisions. Photographed here are four gliders in the landing zone.

Major-General 'Boy' Browning in October 1942 when commanding 1st Airborne Division. By September 1944 he was commander of I British Airborne Corps. (IWM H.24128)

Landing zone 'S' on the first day, with Horsas dropping men and equipment.

C47s dropping men and supply canisters over the drop zone.

Where Louis went into action.

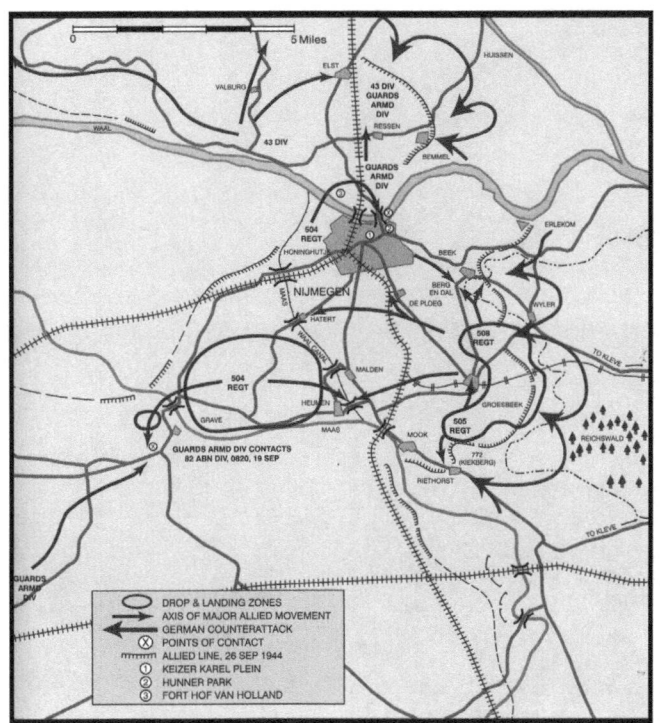

German counter-attacks against the 82nd Airborne Division and XXX Corps.

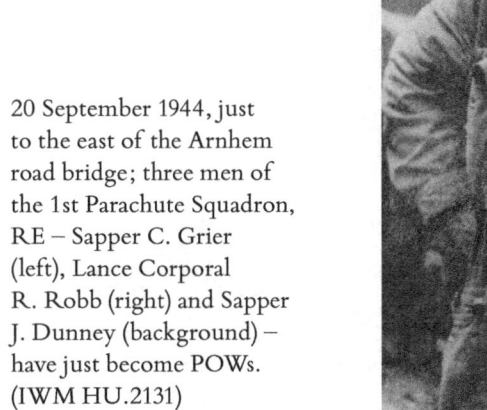

20 September 1944, just to the east of the Arnhem road bridge; three men of the 1st Parachute Squadron, RE – Sapper C. Grier (left), Lance Corporal R. Robb (right) and Sapper J. Dunney (background) – have just become POWs. (IWM HU.2131)

Winrich Behr in 1983.

Louis in later life.

Saturday

It was rather quieter the next morning, and we all felt in a more domestic mood. We cleared up some of the mess in the kitchen; this was indescribable, though obviously unavoidable. We fished hand grenade fuses out of bottled cherries and disentangled small arms and ammunition from the vegetable store. From a pump in one of the back gardens that was not too exposed to sniping we collected water for the day. I did a bit of gardening, which consisted of crawling into the plantation with a sack which I filled with potatoes, tomatoes, sweet corn, fruit and anything handy. In each house there was a large neat stack of onions, and these in themselves made every meal wonderfully tasty.

We had our regular slop-pail stew again. We never got tired of this, there was such a wonderful variety of ingredients. No one ever quite knew what they were eating, as anyone who had collected something just threw it in the pail. Buckley, the only private in our section, slowly developed the job of cook. He resented being asked to cook or being called cook, however, as his real job was that of batman, and he felt he had come down in the world, but when we addressed him like this we meant it as a compliment.

This morning there was even some talk about washing and shaving, but that was going a bit too far as the firing was increasing again and positions had to be carefully manned. Lots of the chaps were pretty sorry sights in the morning after all night in the trenches or firing positions. They had to sit staring into the dark shivering, and worst of all, fighting the irresistible desire to sleep, when they had had hardly any rest at all. Graham and I thought the subsequent session in the cosy Officers' Room was well worth a little patrol every night.

All the men in this sector consisted of senior NCOs who were pilots, like the officers. Such different and separate jobs as officers and men have in the ordinary army did not exist to the same extent in our outfit. We all got on splendidly together and, although we lived for eight days in the most intimate relationship, I never once heard an NCO drop the 'Sir' when addressing an officer. It must be remembered that our present position was unnaturally difficult and somewhat removed from the role we were usually called upon to play on these operations.

If this had not been the case, these pilot officers might never have been tested in leadership and improvisation. The power to improvise and act quickly, in completely unforeseen circumstances, only comes to the fore when the occasion arises.

Only on this comparatively quiet morning did we realise that we were not the only occupants of these houses in the front line. Pale, quiet, frightened people appeared from the cellars. They enquired timidly where they could get some water, and if it would be possible for them to venture into the house proper to collect some blankets and food. There were something like ten people in each cellar, and I went down to see if I could help them, as I was the only one who could make himself understood. I spent the greater part of the morning being asked to various houses to translate their wants and difficulties, although quite a few of them spoke English.

In our house there had lived, unnoticed until now, a grandfather, father, mother and three children, besides some other people who had moved down from a devastated house. None of them complained, and they seemed to be quite apathetic to the destruction of their beautiful houses and furniture.

There appeared to have been a great many Dutch fifth columnists in this area. Most of them were members of the Dutch National Socialist Party, working in the closest cooperation with the Germans, and, in most cases, favouring the total incorporation of their country. The next house to ours belonged to one of these collaborators. The husband had fled with the Germans the moment the first glider appeared in the sky, but the wife and daughter were still in the cellar. The wife had given birth to a boy on the first night of the British occupation. In the rooms of this house we found many photographs of German officers, arm in arm with Dutch girls in uniform, giving the Hitler salute. There was also a gold-framed picture of the owner of the house shaking hands with the Gauleiter. It seemed that the daughter had made friends with some German girls and was carrying on the most idiotic correspondence with them.

We found many German magazines and publications, the weekly war reports were superbly produced; better than anything I have seen here or even from the USA. They really brought the war to the reader – the personal hardships and difficulties of the troops, the supply problems and the strategy. Reading them, it was quite impossible to imagine how Germany could ever lose the war. These publications even impressed us, so their impact on the Germans must have been immense. We got a good laugh, seeing these heroes in print and photograph, when we thought of our personal contact with them.

My study of the German magazines in the Dutch Nazi's house was cut short, as Captain Z and Lieutenant X wanted to take me with them to see the brigadier at Division HQ to report on our patrol of the night before. We kept to the backs of the houses, worming our way from one garden to another. This was not nearly as easy as it should have been, as two of the houses in the row were now occupied by men of the Polish Parachute Brigade, who had come to relieve us the night before. It seemed that these men took war very seriously; and, from reports that had come to us during the night and this morning, great caution had to be exercised whenever we came anywhere near the vicinity of their lines. Captain Z, Lieutenant X and I thought it the best policy to approach them upright and rather noisily. We started talking at the top of our voices and got right into their trench, optimistically thinking we were past the danger, when we were halted by a pair of flashing eyes glaring at us over the top of a rifle. Some sound, which might have been a challenge but was more like a war cry, blared at us from only a yard away, and for the moment we couldn't quite decide what to shout back. Then Lieutenant X managed to think of the password, but it had no effect. The gun was still pointed at us and even Captain Z's hearty 'Don't be a bloody, fool, man!' met with no response at all. It occurred to me that perhaps the kilts worn by the two officers might have put this fierce warrior on the wrong track, so I pushed forward and told him quietly that we wanted to get to Division HQ. Then I tried in German, but he didn't even notice the difference in the two languages. Thank God, one of their officers appeared before the worst could happen. By this time we were in quite a state and Captain Z asked for a Polish guide before we continued our short journey through their lines. Our progress was desperately slow, for whenever we reached another Polish trench or firing position a long and animated argument opened up between our guide and his 'opponent'. Several times we thought we should have to interfere between the two to avert bloodshed. But I suppose it was really only a series of friendly chats between tough guys.

When we got to the large main road, which was under both German and our own fire, it was comparatively easy after what we had just been through. To cross into the perimeter we had to make a dash for it one after the other. Our own trenches were manned and we were greeted in a friendly tongue. They directed us to the Division HQ and to the part where the brigadier was supposed to be at that moment. It was not, as you might think, a respite when we arrived there. The continuous rain of mortars had increased since the morning when I left on patrol and never came back. Everybody who was not working in the cellars of the big house was living and working in very deep, narrow slit-trenches; they looked rather tired and harassed. The mortaring was very uncomfortable and was taking a steady toll on these defenceless people, pinned down as they were.

The nervous strain of it was mirrored in their drawn white faces and unsteady eyes. These, I felt, were the real heroes of Arnhem. It looked very brave to volunteer for patrol and to defend a day position like our row of houses ... but I knew full well that it was not daredevilishness, hate of the Germans or a sense of duty that made me volunteer so readily for the first patrol on Thursday ... but the simple fact that I was afraid of being frightened by the slow rain of mortars.

I soon found that others might have fared better had they acted as I did. Not all had stood up to the strain. In the darkest part of the cellar in the HQ, and in a storage room deep below the gardener's cottage, I found men who had lost all their nerve and self-control. They looked like people who had been seasick for days. Nothing in the world could coax them up. Down there they vegetated; they ate, slept and relieved themselves, in a world where only their fear was reality. They were men, and also a few officers, of all ranks and regiments and of the most varied types. I began to realise how lucky I was. These people in the cellars proved to me that being frightened or not was a matter of luck, like being unmusical or short-sighted.

I knew some of them. There was Sergeant A, quite a close friend of mine back at the station. He was a good-looking chap of 28, always neatly turned out, a good pilot and top of the class in flight theory and navigation. He knew what the war was about and, like all of us, had volunteered to be a glider pilot. He was conscientious, hard-working and unsparing with himself, the kind of person whom I would have thought had all the qualifications of a good soldier. I tried to persuade him to come out of his cellar, for his own good, because he was obviously going through hell, but he simply could not make himself move and, short of using force, there was nothing I could do.

I found Staff Sergeant G in the cellar of the gardener's house. I was even more surprised to see him so terrified and must admit I enjoyed it just for a moment. Staff Sergeant G was the tough guy of our flight ... always getting into scraps with the chaps of other units (usually below his own rank) and called me 'la-di-da' and 'Miss Hagen' because I wore pyjamas and did not swear. He used to brag about what he would do once he met the Germans. He would take the first German child by the legs and tear it in two ... as the only good German was a dead German. There was no sense in reminding him of this because, now that he was so consumed by fear, he was only pathetic.

Seeing so many of our tough guys, including the sergeant-major, in such a terrified state, I realised just why these people in general used such tough language. They were frightened and wanted to give themselves courage and self-confidence. They wanted to appear fierce and fearless and thought that tough and foul language would do the trick. I noticed that those who really stood up to it cursed less the deeper they got into the fighting ... they had proved themselves and did not need encouragement. To hear 'I wonder when the Second Army will come and get us out of this mess' simply stated, without a four-letter word used at least four times, gave me the same surprise as if they had suddenly expressed themselves in Ancient Greek.

Most of the officers and departments were distributed in slit-trenches. There was the Signals HQ trench, the REs' HQ, the Royal Army Service Corps (RASC) people and all the parts of the services which are represented in an airborne army. They all had to have their little offices round the Division HQ and were all working and waiting under this hellish strain. Waiting and waiting for the Second Army. The Second Army was always at the back of our minds. The thought of it made us stand up to anything; not only because we all felt we must hold on and prepare the way for it. That was our job. Everybody was always asking about it, and it was very praiseworthy the way the higher officers patiently answered this eternal question. They just said that, as far as they knew, the Second Army had encountered difficulties behind the lines but they were hoping for its early arrival. We all believed so strongly in it, and as time went on, its strength and power became almost legendary. We liked to imagine the moment when we had armour and could stand up and fight like men.

We found the brigadier walking about regardless of the mortars. He might not even have been conscious of them any more. His job entailed visiting all day, and he couldn't possibly have done this efficiently trying to dodge the slow rain of missiles. Besides the mortars, even inside the perimeter and the Division HQ, sniping was incessant. I must say that the professional German sniper, in his specially spotted cloak, was the most efficient and effective weapon. He could easily have broken the morale of any troops who were not in such a victorious mood as we still were. This may sound contradictory, since we had fought a defensive battle from the moment we arrived, but we knew that we were only a small group in a vast advancing army, and we felt completely superior to the German troops.

The brigadier greeted us and, after Captain Z had told him that I had been on the patrol, laid out his map and asked me to show him exactly where it was I had found the German hideout. He also told me to draw in the movements of the armour I had observed. I was surprised to find that I felt no embarrassment talking to the Officer Commanding the whole Division. I felt I had as much in common with him as anyone in this battle. I proceeded to illustrate our movements on the map, using my grubby finger. It seems that I did the one thing possible to make him lose his temper. 'For crying out loud, take your filthy hand away! You're covering the whole bloody map ... Why don't you get yourself a stick and point it out properly?' And he added, 'They all use their hands; I've had to tell them a hundred times.' 'Yes Sir,' and my interview continued normally.

The brigadier told me he thought it was very interesting, but he must have known better than I did that there was nothing he could do about it. He enquired very precisely the position of the German troops round our street, and said he might be able to get the Second Army artillery from across the river to give us support and ease the situation. He even went so far as to ask me where I thought the barrage would be most effective – in the wood opposite or on the strong-point, right in front. I suggested behind the wood, as the strong-point was really very near our own lines and nothing of importance seemed to be behind it. The main German movement seemed to come from well behind the wood, towards the strong-point. This was farther away from our position, and so gave more margin for the artillery.

Later that night we really did get artillery support from across the river, but of course we couldn't tell how effective it was. We did see fires glowing in that direction, and the explosions sounded strong and very sweet to us. There was beginning to be spasmodic artillery fire from over the river, and the brigadier told us that an artillery officer from the Second Army had arrived at Division HQ and was directing it.

Our own colonel [Murray] now joined the group. He was a tall, striking-looking man, and the thing that everyone noticed about him was his impeccable appearance and perfectly creased trousers. He didn't even wear gaiters. He was the only man who never looked any different from the way he appeared on the parade ground, back at the RAF station. No one has yet found out his recipe for keeping the crease in his trousers. But I found him rather pompous and conceited to talk to, and his get-up, under the circumstances, rather childish. It was he who subsequently organised our whole retreat after the brigadier was wounded. [Brigadier (now General Sir John) Hackett later made a remarkable escape back over the Rhine, many weeks after the battle, with the help of the Dutch resistance.]

These gentlemen were now busy discussing their personal problems of higher strategy, so I retired, after asking Captain Z for permission to stay behind to see the Dutch intelligence officer about the information I had on the subject of Dutch Nazis. Apparently his office was in a trench some distance away. I asked my way from department to department, in other words from trench to trench, and was finally directed to the prisoner-of-war (POW) compound. This had been the tennis court when Division HQ had been a luxury hotel. Now the tennis court was surrounded by a symmetrical zig-zag trench in which were over 200 POWs. Just outside this trench were separate small dug-outs where the German officers were kept. There was one completely isolated dug-out from which projected the steel-helmeted head and broad shoulders of a very impressive figure. He remained quite motionless, like a statue, and seemed oblivious of the mortar shells. He was obviously of very high rank indeed.

I found the POW captain and lieutenant in their trench and was invited to come in and join them in a cup of water. Water was very short in the Division HQ, but the way in which it was offered made refusal impossible, and I was extremely thirsty. I was allowed to ask as many questions as I liked about the prisoners and was made to feel thoroughly at home. My own theory about their quality, and the experience I had had translating for interrogations at the beginning of the action, was confirmed by what I now heard. Except for about a hundred genuine SS troops, there was nothing left of the old arrogance and cockiness. A little later I had a grand opportunity to see this again. The prisoners were shouting and calling for food. Then one of the German officers got up out of his trench. He was quite furious and shouted at them to stop that noise. The British troops, he said sarcastically, had had no food for days and were fighting and disciplined. It was a pity that the Germans had not half the courage and discipline of the British, then they wouldn't have been where they were now. Then our POW officer walked on to the tennis court with his hands in his pockets and stood in the middle with the most disgusted look on his face and called for the parties who had cooked yesterday's dinner.

Our two POW officers couldn't help me in my search for the Dutch intelligence officer [Lieutenant-Commander Arnoldus Wolters, the liaison officer], but they did tell me that he was in naval uniform and usually to be found in the Division HQ building, and this was a step forward. I eventually found him and gave him the information I had discovered, but he told me it was quite impossible to take all the doubtful characters and small Dutch Nazis prisoner as we hadn't the food to feed them or the time to sort them out. He also said that all the real big wigs had gone with the Germans when the first landing occurred on Sunday. He knew of the bother we were having from civilian Dutch snipers, but explained that unless they could actually be caught red-handed, there was nothing to be done about them. On the whole I agreed with him that the population was friendly towards us. I questioned many people, and tried to find out, both in Arnhem and back at Nijmegen, if there had been any organised Dutch resistance groups operating during the German occupation. But as far as we were concerned, there was no sign of them at all – there was definitely nothing here on the lines of the invaluable and splendidly organised French Maquis. [In fact, one of the many tragedies of Arnhem was the British failure to use the Dutch Underground, which was both willing and able.]

There was one large house just outside our perimeter in which there were supposed to be twenty Dutch Nazis. The naval intelligence officer said it might be a good idea for me to collect some men and go and make sure about it. A parachute officer gave me five men, and with them I crossed the main road outside the perimeter. We tried doors, shutters and anything that might give us an entry, but the house seemed to be empty. Then we shot out the lock. It was a lovely old country house, beautifully kept and furnished.

We began our systematic search in the cellars. Here we found the female members of the family huddled together in the dark. They were so terrified of us that there was no doubt about their connection with the Germans. Only when we reached the attic did we find anything that might corroborate our information. A German wireless transmitter with an aerial, still connected to batteries, was there, but there was no sign of any men in the house. We left, taking the wireless with us, after we had destroyed the batteries and aerial. The women in the cellar might easily have been sending information to the Germans, but, as the Dutch officer had said, we would have had to catch them red-handed as in the case of the snipers.

It was high time to get back, as it was lunchtime, and there was no such thing as lunch at Division HQ. I was only just in time when I arrived back. People from the other houses were looking for me, as some civilians had come and they needed an interpreter.

I went to the top house, where they told me that there was a Dutch woman badly injured in one of the unoccupied houses. There seemed to be no sign of life there. I found the cellar door, knocked, and someone opened it. There was one candle shedding a very faint light, and at first I could make out nothing, but as my eyes got accustomed to the gloom I saw that there were eight people in the cellar. There was a jar of water, some odd chairs, a small food supply and some cushions and blankets on the floor. A very pale young woman lay on an improvised bed, and I knew that this must be the injured one. The men were surprisingly neat, in good suits; they were very quiet and courteous and there was no excitement or fuss when they explained what had happened. They pulled back the blankets from the woman's feet and showed me a mess of blood and bandages. She had been shot three days ago, but they had not been able to come out to get help because of the continuous firing. I promised to go over to the hospital and see the MO about her.

I left my Sten gun at the lower house and made a dash across the road to one of our hospital buildings. The entire floor space was covered with casualties. They were all fully dressed and covered with army blankets. I found an officer in the passage, his arm in a sling and his head bandaged, carrying water to the rooms, and asked him for the MO. He told me that the only man who could help me was the RAMC corporal, who was in charge here. The officer found him for me, but the corporal said they were so short-staffed that they couldn't even let me have a medical orderly. He advised me to find the MO [Lieutenant-Colonel Graeme Warrack, the Deputy Director Medical Services], so I had to sprint across the other cross-roads to a different building.

I found the MO in charge, giving orders about fresh water and taking stock of the jeeps still in working order. He was a big cheerful man and said to me, 'You're watching medical history being made, my boy.' He explained that it was quite impossible for him, single-handed as he was, to operate on any of the casualties; all he could do for most of his patients was to smother them with penicillin powder and leave it at that. He also informed me that I was in German territory, as the hospitals were now outside our perimeter. The enemy had taken them two days ago, and he himself, with all its occupants, were prisoners on parole. He pointed out two German soldiers who were the guards of the three buildings; at the moment they were too busy helping with the wounded to do any guarding. Apparently the German MO had arrived and suggested to him that he might like our worst casualties moved behind the German lines. But when the MO told him there were 700 of them he put his hands over his eyes and said, 'Oh mein Gott! That is impossible.' The Germans had very heavy casualties of their own, and, although they had a rear and we had not, they could never have dealt with ours as well.

It was sheer hell for the wounded; they were right in the front line. The German mortar barrage was hitting our perimeter just across the road, twenty-four hours on end. The streets were always swept by our own and German fire, and, until they were knocked out, our 6-pounders fired at German armour approaching along this road. Those men must have felt so terribly helpless lying there, packed like sardines, on every available inch of floor space. The vibration of each explosion made them catch their breath and groan with pain, yet when I went into one of the rooms they all asked me how we were doing and if there was any news of the Second Army.

Obviously the MO couldn't come himself, but he chose an experienced medical orderly to go with me. I showed him down the dark stairs, and he went to work immediately. The first thing he did, after seeing the injury, was to give the woman a morphia injection. Then he began the tedious and revolting process of removing the bandages. The blood had seeped through them and dried; now the dressing was a solid crust all mixed up with what was left of her toes. It took the orderly over an hour. Then he covered her mutilated feet with penicillin powder and left a bottle of this, fresh dressings and morphia with the people, in case he could not get back the next day. During the long and ghastly procedure the inhabitants of the cellar remained calm and quiet. The immediate effect of the morphia put the woman out of pain for the first time for three days. They were all touchingly grateful, though they couldn't say very much. I was glad to be able to tell them that we were only too pleased to do what we could for them, and reminded them that the Dutch were doing wonderful work, helping at the hospital.

I arrived back in our street just in time for the afternoon SP gun attack. Again Lieutenant X and I were able to stop this monster from passing our Piat position. Our attic was quite a cosy little nest, although not as luxurious as the Officers' Room. I suppose it had been a maid's bedroom before it became a gun position. She would have been surprised and shocked to see the use we made of her feather bed. On it was mounted the leg of the Piat and every bomb we fired caused a snowstorm of feathers. I found that these feathers made excellent ear plugs, for my ears were becoming more and more sensitive to the noise of explosions. I appeared at my interview with the brigadier with my ears full of feathers, and only realised later why he and the colonel had looked at me so strangely. The old porcelain washstand in the attic was useful, too, as the basin housed our loose rounds of ammunition and the flat marble top was ideal for laying out and cleaning our Bren. The nicest thing we found up there was a little hoard of food that the maid must have stored away for private snacks. This was really an ideal position from every point of view and we got more and more attached to it.

That afternoon another fleet of supply planes came over to drop urgently needed ammo and food. The cold-blooded pluck and heroism of the pilots was quite incredible. They came in, in their lumbering four-engined machines, at 1,500ft, searching for our position. The ack-ack was such as I have only heard during the worst raids on London, but concentrated on one small area. The German gunners were firing at point-blank range and the supply planes were more or less sitting targets. The rattle of machine guns from the scores of planes, the heavy ack-ack batteries all round us, the sky filled with flashes and puffs of exploding shells, burning planes diving towards the ground, and hundreds and hundreds of red, white, yellow and blue supply parachutes dropping all in this very small area, looked more like an overcrowded and crazy illustration to a child's book. This was war on such a concentrated scale that it made you feel terribly small, frightened and insignificant: something like an ant menaced by a steam roller. All activity on the ground seemed to be suspended and forgotten on both sides. One could do nothing but stare awe-inspired at the inferno above.

How those pilots could have gone into it with their eyes open is beyond my imagination. Later on, when I got back to the 'drome, I heard something of what they had felt. And I was told of their tremendous losses.

When we saw the supply planes coming in over our position, we knew nothing of the hell they had been through already; many of them had failed to get this far. They had first had to deal with great packs of Focke-Wulfs, and in one of the trips they crossed into the Netherlands without any fighter support as the weather did not permit it. When they met the Focke-Wulfs they had very little chance to defend themselves. The Americans were included in our boundless admiration, for they came along in their unarmed, slow, twin-engined Dakotas as regularly as clockwork. The greatest tragedy of all, I think, is that hardly any of these supplies reached us. It makes the heroism of the crews of the planes even more incredible when one realises that they must have known that there was very little chance of their sacrifice being of any use to us.

After the planes left, our usual activities of sniping and covering our approaches continued. We noticed that just across the plantation things were suspiciously quiet. A small patrol was sent over and reported that our troops had withdrawn from these houses. This was highly disturbing as it meant that our street could be surrounded on three sides. The fourth side, nearest to Division HQ, was already under German fire, though not directly threatened. Captain Z sat up in bed and said, 'Bad show, bad show, very dangerous. Wonder what the buggers thought they were up to ... Must see the Colonel about it.' We could easily have moved in now that the row of houses was empty. It was pretty obvious that Jerry would do so if we didn't, as soon as he got up enough courage. Unfortunately for us, nothing was ever done about it; I suppose the reason was that the men needed from Division HQ couldn't be spared. In any case the result was bound to be uncomfortable.

As the afternoon went on, Fearless Frank, one of our glider pilot officers, who had been decorated and earned this name in Sicily, collected some men to go on a patrol. He was a smallish, funny-looking chap, with a terrific guardee moustache that looked absurdly out of place. He had a delicate manner and a gentle voice, which he tried unsuccessfully to turn into a soldierly bark, by emitting his words in clipped splutters, and using a great deal of emphasis and jerky movements of his body.

There were ten of us on this patrol; we were supposed to attack a German-occupied house whose Spandau machine gun was covering the main road running between the two hospitals. Fearless Frank, who was full of initiative and imagination, had quite rightly decided that the only way to deal with this position was to surprise it from the rear – a direction which Jerry would imagine was safely held by his own people. We couldn't really hope to overcome the position, as one side of the house was next to the hospital, the front faced the main road and the other side and the back were in the German lines. But we felt we could give them a jolly good fright, and make it quite uncomfortable for them.

We crossed the small road into a house next to the hospital building, and, from its back garden, into the wood that ran parallel to our street. It was getting dark and we managed to advance without being detected. While advancing we made sure that we had a good line of withdrawal. We cut wire fences and marked trees to direct ourselves. When we got into the back garden of the house we were attacking we could hear the Spandau firing down the road from the front. We knew that our attack must be very short and concentrated – for if they had time to turn the machine gun against us there would be little chance of our getting away.

We put our Brens into position about 20 yards from the back of the house, then simultaneously we opened up with everything we had while we lobbed hand grenades into the windows.

We must have given them quite a fright, for the Spandau stopped and our fire was not returned. The house was completely quiet. It was sickening that this was all we could do; to try and storm the position would have been suicide, and anyway we couldn't have held it. There were only ten of us, with no reinforcements, and Jerryland all round us. We got back without having achieved anything definite, feeling rather vague about what we had done. It wasn't really satisfying.

There were now definite signs that the Germans were moving into the houses across the plantations, just as we had feared. We started to prepare for this new threat. New firing positions had to be fixed up in the back room, the trenches altered so that we could fire in both directions, and bricks taken out of a windowless outhouse in our garden, which made an ideal sniping position. This new situation was very annoying; it meant that we would have to man still more positions and get still less sleep, not to mention the very uncomfortable feeling of being attacked from the rear. We heard Jerry moving about and shouting, but only an odd burst of fire came from this direction during the night. He was firing at us more than ever from the wood, and the large house which stood back in the trees was beginning to catch fire. He must have set it alight purposely, as we had noticed him using incendiary bullets and phosphorous grenades that afternoon. The flames got higher and higher, and we hoped that this would set the whole wood alight and drive Jerry further back from our positions. Unfortunately this never happened and the ruins of the house gave him an ideal advanced sniping position which proved a continual bother to us.

After dinner I was called into the Officers' Room and told that several little jobs had to be done during the night and asked if would I like to be in on them again. A container had been dropped during the afternoon just on the fringe of the wood opposite our houses. We wanted to get it as soon as the flames in the burning house had died down a little. Also, we were nearly out of small-arms ammunition and Piat bombs, and everybody was crying out for cigarettes. We were to try to scrounge whatever we could from Division HQ. This time we were going to make quite sure that the Poles would let us past their lines and Lieutenant S took me to see the Polish officer in charge of the house next door. We got over to their positions about 25 yards away and managed to get into their house by making a dash for it. The Polish officer was most friendly and helpful, and told us that he was going to tell all his men that our patrol was going to pass later on that night. But we wouldn't take any risks this time and we weren't satisfied until he promised us a guide to take us through. When we got back, the fire had died down considerably, and Graham and myself thought it was time to go and recover the container. We could not very well ask anyone to give us covering fire as the road and the position of the container were too exposed and the chances of hitting us or attracting the attention of the enemy were too great. We dashed over and started dragging the container across the road. It was almost impossible to move and, as the burning house was still lighting up this part of the street, everybody must have been able to see us. We had to get across now or never. Thank God the house opposite us was very alert and two pilots came dashing out. The four of us managed to pull the container into the garden. Now that there wasn't such a desperate hurry, we thought it incredible that we had managed to move it at all, because when we tried to drag it further behind the house we could not move it an inch. We opened the container excitedly and found eight wooden boxes; when we broke them open, two large shiny 17-pounder shells were disclosed. One can imagine how pleased we were with our work! If ever there were any 17-pounder guns in our sector they must have been knocked out right in the beginning. We now had the nice job of disposing of them as we could not risk them being blown up by German mortars or shells.

We had a short rest in the Officers' Room and by twelve midnight were on our way to the Division HQ. The patrol consisted of Lieutenant X, Graham, the Polish guide and myself. We crawled along the plantation, very careful not to attract the attention of the new German position, and all went well for a bit. Suddenly, out of the dark, a voice said something

Arnhem Lift

quite unintelligible, sharp and short. It was obviously a Pole challenging us and we looked towards our Polish guide, but he never made a sound. For a moment we thought we had lost him, but I could still feel him next to me lying flat on the ground, his gun pointing towards the voice in the dark; he was trembling. I hissed to him, 'Talk in Polish'. No result whatsoever. We had now taken up the same position behind cover, ready to fire the moment we were attacked. Lieutenant X came crawling back very carefully and poked the Pole in the ribs and whispered, 'Why don't you challenge him in Polish, bloody fool?' But it was not good at all; he remained completely paralysed with fright. The voice in the dark now whispered very fiercely, and it sounded to us like an ultimatum. I started muttering 'British, British' and the password of the day, but the voice seemed to be quite unaware of what I was talking about. We began to move forward now, but the dangerous hissing sound out of the dark made us stop dead. We were in a great state of agitation and I felt mad at our guide. This was a complete deadlock, and the prospect of spending the night there didn't seem very attractive to any of us. We had to try again. Our guide was apparently deaf and dumb, and we pushed him towards the Polish voice, thinking that they might recognise each other when they were near enough. This was a great success. They more or less fell into each other's arms, gibbering like two schoolgirls, and we went into the nearest house, leaving the happy couple behind. We asked for the officer, and after long discussions in sign language, they managed to produce him. Luckily the Polish officer spoke English and realised the danger we were in. He produced one of his men, whom he said could talk English, and we continued our journey. We managed to get past the other Polish-occupied houses with the aid of our new guide, nor was the crossing of the main road to the HQ very hard.

We groped our way to the cellar of the large hotel building. Down there was housed the nerve centre of the Division. Lieutenant X went into the Intelligence Office to report on our position and find out about orders, etc. We waited in the dark gangway. There was an amazing coming and going in the total darkness, everybody tiptoeing and feeling their way quietly, as the floor of the passage was lined on both sides with stretchers on which lay recent casualties. There was a first-aid room with a doctor on one side of this passage, and this was so overcrowded during the night that the wounded could not all be harboured in there. A voice called out 'Anybody know any German here?' I felt my way towards the first-aid room and was asked to interpret for the MO who was attending to a German prisoner. This was a boy of about 18 or 19; his hand and wrist were badly shot up; he looked desperately pale and frightened. He wanted to say something and no one in the room could make out what it was. He said in a small and frightened voice, 'What are they going to do with me now? I was made to fight. It's not my fault.' The doctor made me translate that he shouldn't worry and that he was going to be treated exactly the same as any of our wounded and sent to hospital at the earliest opportunity. His sigh of relief and happy smile made him look quite different. He asked me if we could find his parents when we went into Germany as they had no idea where he was and what had happened to him. I had to take his name and address down, and he was not satisfied until I had done this and given it to the doctor.

Every single German I talked to or watched being interrogated kept on saying that he was made to be a soldier and forced to fight us. It was rather monotonous. The idea of our chaps behaving similarly when taken prisoner by the Germans seemed very funny to me – it was so entirely out of the question. The difference between us and them was that we knew we were right and they knew they were doing wrong and therefore had a guilty conscience. I thought it strange that our German propaganda could not make more use of this tremendous advantage. The thing that most likely stopped them from following this up must have been the policy of unconditional surrender. I wondered if this policy was really worth it.

Lieutenant X came out and said that we could look for the ammo ourselves in the ammunition compound. He had been unsuccessful in getting us any cigarettes, and there was no news of the Second Army, but it still might arrive at any time. It was a painful job looking for the compound. We stumbled into slit-trenches and were pushed out as quickly as we fell in by sleep-drunk soldiers. We were forced to fall flat on our faces time and time again when the shriek of mortars sounded dangerously near. We got there eventually, but what a pathetic ammunition store for a whole division! We could easily have shifted the whole lot into our attic and still had enough space to fire our Piat. Even what there was was not much use, as most of it was 3-inch mortar shells and other kinds of ammo for which there were no weapons.

We did get some ordinary rounds for the Brens and rifles, however, but no Sten ammunition and no 2-inch mortar shells. We couldn't find any Piat bombs either, but Graham remembered having seen some in one of the houses opposite the main road. We made our way back there. He must have had an amazing sense of direction for we found them in complete darkness, hidden under some bushes. Our little expedition had turned out to be quite a successful.

Sunday

At stand-to next morning I was asked about the success of our patrol to Division HQ. What was meant by this was had we brought back any cigarettes? I was considered a complete failure and, to minimise my humiliation, I suggested that they should try smoking the long leaves which hung from the roof of the outdoor building at the back of our house. These were obviously tobacco leaves. Someone cried out, 'Dutch tobacco! Famous the whole world over!' I thought of the terrible disappointment they would feel after they'd smoked the stuff. Their resentment would come back to roost on me. An argument arose about tobacco in general. Someone said that the really good Dutch tobacco came from the colonies, another said that this grown in the Netherlands was the real stuff. Some claimed they had smoked it already. I left them squabbling, knowing that they would find out only too soon. They started to dry the leaves over an open fire and soon everyone was rolling the result in the palms of their hands and stuffing their pipes. The kitchen had to be evacuated almost at once, but some enthusiasts, who were suffering particularly, kept on smoking this filthy concoction till our first cigarette issue in Nijmegen.

The small-arms fire and sniping this morning was worse than anything we had had before. It was doubly effective now. It came from the row of houses across the plantation, as well as from the wood. Our special enemy was the burnt-out house directly opposite us. This was a sniper's paradise and Jerry was using some kind of apparatus to throw hand grenades from behind the house against our façade and barricaded windows. These continuous explosions right under our noses were really very uncomfortable. Our barricades of the windows and doors had to be redone every time the blast of a grenade threw them back into the room. There is nothing more irritating than a grenade bounding into a furnished room and rolling perhaps under the bed, where it can't be fished out and thrown back before it explodes.

The solution to our troubles would have been simply to knock the building flat with mortars or our Piat gun, but we had no mortar bombs at all, and the Piat bombs we had collected could not be spared for a job of this sort.

I changed my Bren-gun position from the attic into the front room of the second floor. I started to give them bursts any time I could detect the slightest movement. Smithy was beside me, sniping with his rifle through a hole in the wall near the window. The crackling of small arms fire was like a bonfire. The Germans had probably detected our firing positions. We could hear the thud of the bullets on the outside wall and the shape of each window was outlined and filled in by a pattern of bullet holes on the wall behind us. Each time they hit the edge of the windows a spray of chips, splinters and plaster made us jump aside. They seemed to be giving Smithy and me all their attention.

Suddenly Smithy shouted, 'I've been hit.' His whole wrist and hand were soaked in blood. He lay on the floor while I tried to tie a bit of rag round the wound to stop the bleeding. Then we both crawled across the room to the door and I shouted for someone to help him down the stairs into the Officers' Room. I had to stay up there as the Bren was too valuable for our defence to be left inactive.

I poked my barrel through the hole that Smithy had left and started firing at some movement I detected. Then a sudden terrific bang ... I thought this was my turn. My hand was covered with blood and I withdrew the gun quickly. I wiped the blood away, expecting to find a serious injury as I could feel no pain at all, and discovered that a splinter had just penetrated a vein and this was causing the bleeding. And then I noticed my Bren barrel and realised how incredibly lucky I had been. A bullet had entered the flash-eliminator that widens the barrel at the end, split it open and by some miracle ricocheted off again, instead of going into me. The good old gun still worked, but I had to exchange barrels at the earliest opportunity. Anyhow, after a few minutes the signal for my morning session in the attic was heard from the top of the street, and, calling to Lieutenant X, I fetched the Piat and went up the stairs, quite confident, and taking my time. By now we knew the slow and careful approach of our old friend, the German SP gun.

This morning, though, they had a surprise for us. They manoeuvred two of them into position, one on each side of the road. The terrific small-arms attack seemed to show that they were working up for real business. We were glad we had fetched those twenty bombs last night, as with two SPs firing shells incessantly as they advanced, we had to lay a screen across the road. This was our only hope of stopping them. They stopped dead for a while, just out of range, and then retired slowly. The small-arms activity died down to normal and we knew we had repelled them again, anyhow for the moment.

It suddenly became a pleasant Sunday morning. Our regular Arnhem hotpot was simmering, the Red Cross appeared on the streets, and everyone came down for a breather to see how the food was getting on and how the others were. Eating our food, the reaction from the heavy attack made us quite jolly and we compared our near misses. Everybody boasted and produced proof. I thought I had a pretty good story to tell, but when I saw some of the others I knew that this morning's battle honours were not mine. Like all of us, Tony wore a sort of false bosom under his battle smock. This consisted of his personal belongings and ammunition. A bullet has passed through this bosom, into his water bottle, smashed up the pin of a hand grenade in his top pocket and passed out again. His first reaction had been fear that he was mortally wounded, for water from his bottle was trickling down his body and he naturally thought this must be blood. After examining himself, he removed all his belongings and got an even worse fright when he pulled out a live grenade minus the pin. He had thrown it away before he realised what he was doing, and the fright (though of course he said it was the explosion) knocked him for six. Anybody who survived an experience like that deserved to get out alive, and Tony did.

The next of this morning's heroes was Fitz. His private bullet, which he now thinks of with great affection, entered his false bosom and ripped his smock and the tunic beneath it three-quarters of the way across his chest, just above his heart. There were many minor escapees like myself with my Bren gun. I fetched this down proudly and passed it round the kitchen. Then Vic discovered that my left epaulette had been split across by a bullet. I felt I could not join the other two on equal terms. As a matter of fact, none of us could hold a candle to our private miracle man. This glider pilot was the pride of our street, because by all the laws of nature he ought to have been dead, and instead he was doing kitchen fatigues quite happily.

A bullet had entered his right temple and exited through his left, leaving behind it a couple of neat little holes. He wasn't even knocked out of the fight by this and had to be ordered sternly not to take part in combatant duties. Not only did he remain working in the kitchen all of the time we were in our house, but he was able to retreat with us across the Rhine. The last we saw of him he was sitting perkily in the back of an ambulance in Belgium, making rude signs at us as he passed. An MO whom we told about it later, said that this was possible, as the front part of the brain governs the emotions, and an injury to it does not necessarily cause any organic change. He explained that the work done by the damaged part of the brain would be taken on temporarily by another part. And he added that quite often people injured in this way become very cheerful to start with and feel stimulated. This, of course, does not last for long, but would account for the way our miracle man behaved.

Sunday afternoon was fairly quiet, but towards dusk the firing increased sharply. Before we knew what was happening it developed into the first direct assault on our position. Somehow Jerry had crossed the road from the wood into the next house. I hadn't seen it happen, so they might have come from the plantation in our rear. The first thing I saw were the tops of German helmets moving along the space between the two houses, which only gave them very slight cover. I had to get a firing line from my attic to them, which meant removing several tiles, and by the time I got into position they were firing at us with an automatic gun. I could take my time aiming and getting ready as it was very unlikely that they would detect me. After my first burst the German who was firing disappeared and his gun toppled over on to the veranda. Either myself, or someone else firing in the same direction, had hit one or more of them. I started looking round for new victims. It was getting dark now and difficult to make out what was going on. Right underneath me next door, only about 3 yards away, a window was pushed open and I saw people moving inside the room. So the Germans were here! This room had always been empty and lately the whole house had been evacuated to strengthen the top corner position. They were trying to get out of the window into our house and I could not fire at them from where I was as they were too close underneath me. I ran from my attic to a side window on the second floor to lob a hand grenade into the room. I heard it break some glass and a few seconds later it exploded. When my eyes got used to the darkness again, I could still see movement in the room after it had exploded. I couldn't make out why I had not killed everyone in there and threw, one after another, my remaining store of grenades. And still there was movement in the room. I was just going to run down and fetch some more hand grenades to deal with this apparent wave of Germans when a controlled and quiet voice called up to me, 'What do you think you are doing?' Trying to kill us all?' My heart stood still for a moment. Then I realised what I had been doing. I was certain I had killed and injured many of them; it was the worst moment of the whole seven days, and I wished I was dead myself. I jumped down the stairs out of the house and across to the other one through the window and into the room. Everything looked quite normal there, except for the miracle man lying on the bed with a bandage round his head and the lieutenant and three others sitting on the floor. I told them that it was I who had thrown the grenades, and where were the casualties? He said, 'Oh, it was you, was it? Thank God you didn't know your job. At such short distance you should have waited four seconds until you threw; that gives another three seconds until the grenade explodes. As it was, we lobbed them out of the window as fast as you threw them. They all exploded just outside, and you're a fool not to have noticed it.' I was never more grateful for being a fool!

The others had been fighting hard all round the house, as the enemy had tried to get at us from every side. It began to get quieter and it seemed as if they had had enough. Next morning we found several bodies, amongst them my Jerry machine-gunner.

I was called to the Officers' Room and Captain Z told me to wake him at 1 a.m. as he had to report to the Brigadier at 2 a.m. I was to go with him to Division HQ, so I made my bed, and Lieutenant X, who was duty officer for that period, promised to wake me in time. It was hell to get up at this hour of the morning for I had only had two hours' sleep. The light was dim and it was peaceful and warm. Captain Z's full rosy cheeks, and the regular vibrations of the ends of his moustache, made a picture of peace and content. I could hardly bring myself to wake him. I shook him gently, but soon realised that sterner measures would have to be used. I pulled the sheets and eiderdown off him, shook him roughly and shouted at him, 'Captain Z, you have to get up to go to the Div. HQ.' 'Who's giving orders here?' he replied. 'The Brigadier, sir, and it's nearly half-past one.'

We arrived in the cellar at Division HQ; the little room of the Intelligence people was very crowded and no one took any notice of us. Everyone was half asleep and could hardly keep their eyes open. I reported the glider pilot section under Captain Z present and pointed out on the map our position, and that of the enemy. There were no orders for the next day, except to carry on as usual, and a bit of encouragement by telling us that we had been doing well. We went on another scrounge and returned with a few Piat bombs and several hundred rounds of rifle ammo.

Monday

It was nearly dawn on Monday morning by the time we got back, so we helped the duty officer and the sergeant-major to wake everybody and get ready for the stand-to. The firing slowly increased from all sides, but we did not have to cope with any strong attacks. I think we must have inflicted fairly heavy casualties yesterday evening and so discouraged Jerry from another assault. We were all desperately tired by now and everyone seemed to be sleep-walking and dozing off every now and then at their firing positions. I felt like a very slow and heavy machine, doing everything automatically, but at half speed. The whole morning was quieter than it had ever been before and the lack of action made us feel our weariness. The SP guns did not even appear during the morning. There seemed to be a complete deadlock at our sector. Probably Jerry was preparing a big attack. We were hoping for the Second Army. This morning gave us a chance to think and we realised that we could not hold on for ever without being reinforced. Lunch cheered us up again, and I quite enjoyed hearing an increase in the German small-arms fire, as it meant we had to fire back, and that was the only way to keep awake.

Captain Z and another captain came back from Division HQ and the officers from the other houses came filing through our kitchen into the Officers' Room. Something was up. Rumours began to circulate at once. The Second Army was on its way; we were going to withdraw; we were going to attack; our street was going to be withdrawn into the perimeter; and so on and so on. The officers' huddle went on for two hours and the rumours went on circulating.

At last we were told to go and see the captain in groups of four for the night's orders. I went in. A large map was spread on the table. In a confident voice, Captain Z began the briefing. We were going to retreat across the Rhine and join the Second Army. It was going to be an orderly and organised withdrawal. All sections were to leave their sectors at a specified time with all their arms and ammunition. Our street was going to leave at ten-fifteen. The men from the top corner house would move in to the next house; from there they would all move into the next, and so on until the entire glider pilot section were all in the house nearest the perimeter. From there he pointed out the route that we were going to take to the river. It lay mostly through woods and along little paths, and the more we tried to memorise it the harder we knew it was going to be. All except Captain Z. He fairly radiated optimism.

When I came out of the Officers' Room into the kitchen, quite a few of the chaps were stuffing themselves with food, trying to lay up a store of energy for whatever might be coming. It seemed criminal to be leaving any food behind and no one had any idea when their next meal was going to be. I thought it very wise and joined the party. Lieutenant X and two other men worked their way through the shadowy kitchen carrying a stretcher. Bill had been killed the night before and had to be buried. We stood up, and there was a minute's silence as they passed out through the back door, then we went on with our supper.

I got my gun ready and clean, collected odd Sten-gun rounds from all the rooms in the house, and filled the empty mags. I was hoping for two hours' rest, but firing had increased so much that we had to go out and man all the positions. We didn't want to run any risks at this particular moment and we wanted to give an impression of complete normality. We had been specially warned at the briefing not even to discuss the coming withdrawal among ourselves, for fear of listening Germans or civilians. Any leakage would have resulted in a bloody slaughter. All our boots were to be covered in sacking and those who got hold of civilian shoes were to hang their boots round their necks. We were supposed to blacken our faces, but this was hardly necessary, with a week's growth of beard and no washing. The order for the whole Division was to withdraw in single file, fully armed and without noise.

It may have been nerves or over-anxiety to appear normal, but all along our street our positions started blazing against the Jerry lines. I was pleased to notice that his reply was not over-energetic. I stood with Vic and Cooper in the trench, peering out into the darkness and burying 3-inch mortar shells which we had never been able to use during the whole action. I was leaving the trench to try and find a spade when a burst of fire came straight at us from the next house. Someone said, 'They've hit me,' and we found Cooper leaning against the wall holding his right arm. We took him up to the kitchen and then dashed back to the trench. We knew it must have been someone from the next house and went down the trench towards it. There was no one on guard there and we entered the side door. The entrance was crowded with chattering Polish soldiers. Before I could find their officer, one of them had elbowed his way through and asked me how many of us were killed or injured. I told him that we had been very lucky and his men had only hit one of our chaps. A soldier came up to me, talking very excitedly with tears in his eyes. Someone came and translated. He wanted to know if he had killed or seriously injured anyone, and was terribly relieved when he heard that all four bullets had gone into the arm of only one man. He went on to explain that he thought the Germans had infiltrated the trench and, when there was no reply to his challenge, he fired some rounds. There was nothing one could do; but it was specially sad, as it was nine o'clock now and it would give Cooper very little chance to recover in time for the withdrawal. Besides, orders were not to take any injured men with us; but, in actual fact, Cooper stood up to the strain of the withdrawal and I saw him safely in a hospital with the Second Army.

It was nearly zero hour. Captain Z called me to the Officers' Room and told me to stick to him all through the withdrawal. More and more men from the other houses were coming down to our position, waiting for Captain Z to lead them to safety. We moved off, Captain Z and myself leading. Behind us a silent, long file of about fifty glider pilots. We made our way steadily to Division HQ. It was easy for us as we had been this way every night. We wound our way through the mass of slit-trenches, trying to avoid any obstacle which might break the long chain. It was a black night, and the noise of the dripping from the trees covered the sound of our footsteps. We passed within 50 yards of the POW compound. Up to now we hadn't met a single soul and even when I looked over to the old tennis court I could not see any movement whatsoever. Firing ahead of us could be heard quite distinctly, mostly bursts of machine gun and the thud of exploding mortars. We were leaving our old perimeter now and moving through a kind of no-man's-land towards the river. The Germans had not been able to occupy this part thoroughly, but machine-gun nests and strong-points were dotted about in the woods. It was about 4 miles from here to the river and the problem was to get there, avoiding these danger spots.

Captain Z still seemed pretty sure of the route he was taking, but the denser the woods and undergrowth became, the more difficult it was to follow the path. In front of us was a large meadow and we had to find a way round it; we could not risk crossing it openly. From then onwards it was more or less intuition and the colossal luck of Captain Z that brought us to our destination. He felt his way forward, muttering to me, 'We'll make it yet, don't worry, you stick to me ... Do you think we are all right, old boy?' I didn't have the slightest idea whether we were all right or where we were, nor had anyone else. I felt rather like the blind leading the blind. Miraculously, we came upon a farmhouse which we recognised as having been marked on the map. We knew we had to take the lane to the left of this and we followed it. Machine guns could be heard ahead of us and, as our column halted to climb over a fence, a breathless officer appeared out of the darkness and told us to turn round immediately as his column had run straight into a German Spandau and he thought he was the only one who had survived. His head was bleeding and he had tied a bandage round it himself. We turned round, everyone following the man in front, until the leading part of the column had overtaken the tail. The men were remarkably silent and disciplined and there was no shuffling or pushing. Captain Z asked me to go along the line and call all the men with Sten guns to the front. This was the most dangerous part of our journey; we might run into the enemy at any moment. On and on we walked through the dark wood, turning and winding whenever we felt we were too near the enemy. I was walking in front of Captain Z, my finger on the trigger, prepared for anything. I had lost all sense of time or distance and groped my way forward wherever he directed me. We emerged from the wood and in front of us stretched a wide plain. This was the approach to the river. Someone came up and led us along the fringe of the wood. We followed him until we came to a white tape which was stretched across the plain. We were to follow this until we reached the river. 'We've done it again, we've done it again!' Captain Z whispered excitedly to me as the long column followed the seemingly endless white tape across the meadows. From the wood behind us firing could be heard and we wondered if they had found us out. Mortar shells were passing over, exploding not very far in front. The crossing was not going to be easy; we knew that by now.

The tape led us to a hedge and continued alongside it, eventually passing through to the other side, on to a path running down towards the river. The mortar barrage must have shifted, or even stopped, and the relief at the thought of a safe crossing was wonderful. Captain Z said, 'Look, the poor dead cow.' Then we began to see human bodies lying all along the path. As we continued, we saw that some of them were moving and heard groans and weak cries for help. A voice in the dark was muttering and talking in delirium. This was more than I could stand and I knelt down by one of the injured men. I called for three of our chaps to help me take him down to the river. The wounded man begged us not to leave him behind and to help him across the river. We tried to lift him, but he groaned with pain and we had to lay him down again.

Then as my eyes got used to the open darkness of the meadow, searching for the origin of subdued screams, I began to distinguish the shapes of bodies dragging themselves towards the path. Feverish pleading eyes looked up towards me, arms clutched around my legs, it seemed that all the wounded were frenzied by the fear of being left behind. For the first time during the whole action I panicked. I dragged limp bodies along towards the beach. I ran around in circles searching for someone in command and pleading with uninjured men to give me a hand. I vomited and felt faint. Then someone with an authoritative voice came up from the river and ordered me to leave the wounded where they were as they could not be got over the river just now and a doctor would be left behind to look after them. Exhausted and dazed by my impotence and the ghastliness of the scene, I continued towards the river.

All along the path there were mortar pits and the bodies of dead and wounded soldiers. We reached the banks of the Rhine and joined a long queue of men waiting to be ferried across. Someone came up to us and told us to spread out as the mortaring might be resumed any minute. There were at least a hundred men in front of us and no sign of a boat. There were other parties like ours all along the river, waiting. The splash of oars could be heard now and then. I suppose this was how they felt at Dunkirk. A small canvas boat was approaching at last. It took ten men across. Then we realised our desperate position. Any moment the mortaring might start again. There was no cover at all and we crouched in the deep squelchy mud. We were frozen with cold and soaked from the rain.

The mortaring started up again, not directly where we were, but near enough to be frightening. After trenches and street fighting, and even the cover of the woods, we felt helplessly exposed. The thought of those ghastly bodies and the groans of the wounded, lying in the meadows, was in everyone's mind, but no one said anything. We just crouched there shivering.

I began surveying our position in my mind. Of course this had nothing in common with Dunkirk, and those who ordered us to wait in line patiently until we were taken off by those ridiculous little canvas assault boats did not know what they were doing. The Rhine was only 250 yards wide and quite narrow at certain spots near us. Why was the order not given for those of us who could swim to dump their arms and make for the other side? Surely it would have been possible to organise a rope and stretch it across for those who were not strong swimmers? But instead we were being heroic, playing at Dunkirk, and a great many men who could have escaped to safety would be casualties or else be taken prisoner at dawn.

I had to get out of this. I told Captain Z that I couldn't stand this any longer and that I was going to try and swim for it. Now we had got this far I didn't intend to take any more risks than were necessary. The boat system was obviously hopelessly inadequate and, apart from relieving some of this awful congestion on the bank and leaving the boats, such as they were, to the non-swimmers, I honestly thought it was the best way out. He agreed with me and shouted to the rest of our glider pilot section that we were going on to a promontory where the river narrowed a bit.

A large crowd followed us, but I doubt if any of them realised where we were going or what we intended to do. They just came after us because at least we seemed to have some kind of plan. Had they been told that the river was only 250 yards wide, though it looked rather more in the dark, many would have followed us, orders or no orders. We had to climb some large boulders on our way to the promontory. At the end, it went steeply down into the water and would have made a far better landing stage for the rescue boats than the mud flats, as at least the bank gave a little cover. From here the opposite bank didn't look too far and the prospect of doing something after the misery of queuing up on all fours in the mud made Captain Z and me feel quite cheerful. 'We'll do it again, you and me!' he said. We proceeded to take our boots off and hung them round our necks. Captain Z gave his rifle to Lieutenant X, who unfortunately couldn't swim, and remarked that he must keep his haversack with him as the Company 'Office' etc., was in it. I kept all my arms and ammo as we couldn't be sure what would greet us on the other side.

I put my Sten gun across my shoulders and, by the time I was ready, Captain Z was in the water and swimming away from the bank. In I went. The water was pleasantly warm, the air filling my battle smock kept me easily afloat. I felt happy and full of confidence. Captain Z was about 20 yards in front of me, but drifting fast downstream. The current was very strong and I tried hard to work against it so as to reach the other bank more or less opposite where the promontory stood. Captain Z seemed to be getting on all right and I couldn't catch him up, as my battle smock gradually deflated and swimming became harder and harder. I began to get worried and breathe fast and I was only halfway across. Then I wasn't doing proper strokes any more and I began to panic. Like a flash it came to me that this was the one fatal thing to do and the best possible way to get drowned. How ridiculous it would seem that I, brought up as I had been by a lake, and a swimmer since I was 4, should die by drowning in the calm warm waters of the Rhine after evading every kind of violent death for the last seven days.

I turned over on my back to rest and pull myself together. I realised that I had to get rid of my Sten gun, but that would be pretty difficult as it was strapped round my back. I had to let myself sink vertically while I eased the gun up and over my head. A moment later I heard it go bubbling to the bottom. Next, I methodically rid myself of all the impediments that my battle smock contained, also my boots and steel helmet. There were Sten-gun mags, hand grenades, writing materials, my fountain pen and every conceivable thing I had managed to save. Unfortunately I couldn't discriminate in the water, and all my belongings, including my AB 64 [identity papers], went floating down the Rhine.

The difference was marvellous. I felt like I had when I'd been bathing a fortnight ago in the Thames, except that it was dark. I looked round for Captain Z, but there was no sign of him at all. I shouted and began to swim round, but there was no reply and I supposed he had already got to the far bank. I swam on alone. Fires were burning on both sides of the river and the mortaring was still going on. There was an increased rattle of machine guns from the wood we had left only an hour ago, but I felt I was beyond this, enclosed by the still warm waters. I might never have been on the other side.

I was about 20 yards from land when I saw two figures gesticulating wildly and heard shouting: 'Hold on, mate, hold on. We'll be there in a moment. Don't panic, it's OK, you're safe now.' They were preparing to plunge into the water and pull me out when I shouted to them not to bother and that I was perfectly all right. Two pairs of hands seized me and pulled me out. Nothing would convince these enthusiastic life-savers that I didn't need artificial respiration. Maybe they'd been on a course some time and this was their unique chance to practise the real thing. They tried to turn me upside down, but only succeeded in pushing me face downwards into the mud. They wouldn't let me walk on my own, but tried to lift me. This was quite impossible on the slippery ground, and anyway I would probably have been more able to carry them as they were rather undersized.

I did all I could to persuade them to let me go on alone while they stayed behind to watch for other swimmers, especially Captain Z. I began to be very worried about him when they told me that no one had come across this way and I was the first they had 'saved'. But they insisted on coming with me.

I was shivering and slipping about on the muddy ground, but what the hell? I was here and so was the Second Army. We slid and stumbled along for about fifteen minutes until we met a medical orderly. The two cockneys instructed him to take me to the first-aid post and not to stand any nonsense if I tried to escape. They said in very important voices, 'He swum the Rhine and we fished him out.' As far as they were concerned I was still drowned.

We joined a long stream of shivering men walking to the main road. Then we turned left to an open space where some lorries were parked. These were supposed to be for the wounded, but the orderly bundled me in and, as I was terribly cold and weary and had no boots, I didn't protest too loudly. We drove along without lights, as this side of the river was also under enemy fire. Each time the lorry bumped round the edge of a shell hole the injured groaned and swore.

We stopped in front of the tents of a first-aid post and were helped out and taken inside by medical orderlies. A queue was slowly filing past several tables on which were assembled bandages, syringes, bottles and files, etc. MOs and orderlies were treating the walking casualties with lightning speed and efficiency. There was nothing for me to do but join the queue, as particulars were being taken and instructions given by a clerk sitting at the last table of all. I reached the first table shivering more than ever and was passed on without comment to an orderly who painted something on my forehead and was about to give me a morphia injection when I protested energetically. He replied curtly, 'I'm treating for shock here; you'll feel fine in a minute.' He went on with the treatment until I had to resort to physical violence. This made things worse; he now thought I was a really bad shock case with a touch of neurosis thrown in. He fetched a doctor and I explained that there was absolutely nothing wrong with me, except that I was very cold and wet. With the help of another orderly I undressed and my wet uniform was thrown on a large pile of soaked and bloody garments, and that was the last I saw of it. They wrapped me in blankets, gave me a cup of very sweet tea, sat me on a chair and lit a little oil stove underneath me. Then they put a cigarette in my mouth. I was in paradise.

Several ambulances were waiting outside; they were for the walking wounded, to take them on the next stage of their journey, en route for Nijmegen. I climbed into a very crowded one and, at the next stop, managed to join a group of soldiers who were going to Nijmegen, where everyone who had escaped was supposed to report. I was still barefoot and draped in blankets, like a Roman emperor, but found that there was no need to be self-conscious, as I was by no means the only one improperly dressed. We piled out of the lorries and filed into a large hall. A sort of improvised banquet was awaiting us; there were candles on the tables, wine glasses full of rum, and the most delicious and enormous three-course supper, rounded off with the inevitable cup of tea. Our headquarters staff, including cooks, quartermaster's staff, etc., had been waiting here to join us in Arnhem, along with the Second Army. It was they who prepared this wonderful reception for us. As we left the hall we gave our names and numbers to our squadron clerks and were given a ticket bearing the room number and barracks where we were to sleep. It was all very touching and efficient.

I spent the next twelve hours sleeping dreamlessly in a beautifully made bed.

When I woke up it occurred to me that I couldn't walk about Nijmegen still dressed as a Roman emperor, as our barracks was right in the middle of the town. Our own quartermaster stores lorry, filled with most necessities, was waiting outside and I was able to procure enough garments to cover my nakedness. The rest I scrounged from all sorts of other stores and places.

A parade was called and many of the officers and men who had only arrived during the morning, and had had to walk all the way from the river, had to be collected from wherever they were. The roll was called by an officer in the same sort of undress as I had been when I arrived. He called each name and, if the man were not present, anyone who knew his whereabouts or fate had to speak up. We stood outside the barracks, near the air-raid shelters in which hundreds of Dutch civilians were living. There was no air raid or shelling just now, so they collected round us to see the spectacle. Every time the officer had to write down a report of a missing man, his blanket had a strong tendency to slip down, and everyone watched, fascinated. From time to time it did fall right down before anyone could warn him. The civilian women looked away modestly. He didn't seem to take any notice as his job was far too important for him to worry about a detail like that.

The lovely dining hall where we had eaten the night before received a direct hit during the morning, so we had to queue up at the cookhouse for our plentiful lunch. By now we were getting used to civilisation again and were thinking how good it would be to sleep and live in one of these charming and spotless Dutch houses surrounding the barracks. Vic and I decided to go on patrol – we had got the habit. We didn't venture very far, but went to a house just opposite, where I asked the *Mevrouw* if she could lend us a needle and thread as my poor friend had torn his trousers. She was only too delighted to do the job herself and Vic sat comfortably in *Mijnheer*'s easy chair and dressing gown. Eventually we cadged an invitation to stay the night there and use the house as if it were our own. This was just what we wanted, and most satisfactory. We wound up our stay in Nijmegen with an Anglo-Dutch party. We ate scrounged Second Army rations and drank Dutch cognac.

Before we left, General Browning called the remnants of the Division together and made an impressive and very sensible speech. We were in no mood to hear about glory and battle honours and the thanks of the motherland. We had lost, or had to leave behind, too many of our friends. But we were interested to know if the sacrifice of this Airborne Division [1st Airborne Division] would prove to have been of any use, and why the Second Army had not been able to join up with us. General Browning told us quite straightforwardly what we wanted to know. We realised what he must have felt, waiting with the Second Army, and knowing better than we ever did just how desperate our position was.

We were glad to leave Nijmegen, as what we were all longing for was a spell of complete safety. We simply loathed the air raids and spent our time diving into shelters along with the civilians, and once we had slept off our first weariness we didn't care for sleeping on the top floors of the barracks. A London civilian would have put us to shame. The idea of being killed in an air raid, after surviving all the dangers of the past days, was more than anyone could stand.

On Thursday morning the whole Division left in one huge convoy along the main Nijmegen–Brussels road. The civilians, who had been most friendly and hospitable, came searching through the crowd of waiting soldiers to say goodbye to their special friends. We departed with souvenirs and gifts. All the way to Louvain we passed one continuous stream of transport, tanks, artillery, petrol lorries, jeeps, ducks [DUKWs – amphibious vehicles] and RAF vehicles. The Second Army was every bit as strong as we had imagined it. We knew that it would be able to finish the job we had had to abandon.

In parts of the road, where the corridor was especially narrow, we were still shelled by German artillery, and British tanks were covering the cross-roads. Right down to Brussels the road was lined with burnt-out Allied and German transport, armoured cars and tanks. Here and there were groups of crosses with German or British steel helmets on them. There were villages completely devastated, and scores of burnt-out tanks clustered together. These must have been the scenes of German breakthroughs into the corridor, and the eventual defeat of these columns.

Everything was prepared for us in Louvain and we lined up for our supper, tea and dinner combined as soon as we arrived in the late afternoon. There weren't any air raids, but the lovely town and library were very badly damaged. We spent the evening pub-crawling and making friends with the Belgians. We drank lots of cognac. Early next morning we visited the vegetable market. It was overflowing with fruit and the most delicious grapes. The Belgian peasants would hardly allow us to pay for any of our purchases. It was surprising how enthusiastic they still were; after all, we weren't the first Allied troops to enter the city [nearby Brussels had been liberated on 3 September].

Later in the morning we were taken to an aerodrome and embarked for England. I slept the whole way across and so can give no account of a reaction – tears in my eyes or lump in my throat – to the sight of the White Cliffs. I suppose that makes this story rather incomplete.

By the evening I was back at my own 'drome. I had been away twelve days. It seemed like one day, or a lifetime.

Our huts had been locked and left just as they were when we left them on Monday morning. We took the keys, went in and sat down on our beds. The four of us looked round the hut. There were eighteen empty beds. It was very quiet now and we remembered the noise and bustle there had always been before we went away.

Supper was waiting for us in the mess. Tonight we didn't have to queue up, we were waited on by RAF pilots and WAAF. A party of all ranks was in full swing in the NAAFI. This was for us. They had expected us much earlier and a cordon of strong men was thrown round the two remaining beer barrels; but all the same our hosts were already well away and were sending sorties of shock troops to break the cordon. An RAF pilot officer was helped on to a table to deliver a speech of welcome. It wasn't long nor was it very profound, but it was received with an ear-splitting 'Hip-Hip-Hooray'. Then we were called on to answer it.

A glider pilot was carried on to the table. He spoke rather well, more coherently than the pilot officer – this was understandable. He said how much we had appreciated the gallant efforts of the RAF to reinforce us; this had struck him as the most heroic feat of the whole action. He thanked the whole ground staff for the magnificent work they had done, getting the gliders and tugs ready and in perfect trim. A terrific cheer from the glider pilots greeted this, and the party went on.

When I got to bed after the party I didn't fall asleep immediately. Perhaps it was the atmosphere of the nearly empty hut. I began to think backwards for the first time. Odd things occurred to me, not particularly important things, and in no special order. The life we had led at Arnhem was nearer to animal existence than anything we could have conceived, and yet the more savage the fighting got, the more civilised the men seemed to become. By civilised I don't mean having baths and being clean and shaving and eating with a knife and fork, but the relations between man and man. They became increasingly polite and helpful.

There was such gentleness and friendship among them as would have made any of them almost uncomfortable back on the station. Although they were fighting like tigers, and in that fight had to be completely ruthless, there was no tough behaviour or coarseness of speech. It was almost uncanny. The familiar army swear words and idioms were absent from their conversation, probably for the first time since any of them joined the service. They were courteous, kind and considerate, without any self-consciousness. I remembered the awful moment when I had had to admit that it was I who had thrown those grenades into the next house. Ten days before, if I had trodden on the toe of one of those men, a stream of filthy abuse would have been hurled at me. Now, all they did was to point out politely and with no recriminations, that I wasn't very clever. I remembered Cooper being shot by the Pole, and the quiet way he and the others took it and even felt sorry for the weeping soldier who had wounded him. Then the withdrawal in single file, no pushing or jostling; and the complete self-control of the men crouching in the mud, waiting for the boats; the way they passed the lightly wounded to the front. Their concern for the Dutch civilians, and the complete absence of grumbling and bitterness. That is what I call civilised behaviour.

Then I remembered some of the German troops at Arnhem; the continuous shouting and whining, which, even if one hadn't understood it, gave the impression of savagery. Their lack of self-discipline, their desire to get out of everything and not be the one to carry the baby: 'Why send me? Let him do it' or 'Do it yourself.' From my attic, I actually saw a German officer take off his hat and stamp on it in desperation and rage, when his men were quibbling about who should go forward against our positions. I later realised that these were non-combatant troops rounded up to serve in an emergency.

I began to think about what we had talked of among ourselves when we weren't actually fighting. How we had admitted without shame when we were frightened, telling each other about our feelings, recounting the incredible deeds of others. Under ordinary conditions, soldiers never stop talking sex: 'Subject normal.' During the whole seven days I was there, it just ceased to be a topic of conversation or enter our minds.

Nor did anyone mention home and family. Perhaps this was because we thought that no one knew of our desperate position and there was no need to worry about them worrying. I really don't think any of us thought at all; we were too busy living and we seemed to act almost entirely by instinct. None of us will probably ever be so natural again as we were there. We were completely without inhibitions; there wasn't time for them.

Looking back I realised that I now had a complete picture of all the people who had been there with me. Then I knew that I had a complete picture of myself. The seven days had given me seven years of experience and confidence; I knew what I was like … Then I went to sleep.

4

Winrich Behr's Story

While I was in Germany in the summer of 1990 I was invited to a dinner party by a distant relative, Grafin Nona Oeynhausen-Sierstorff, at her country house. I was seated next to a good-looking grey-haired man of roughly my own age. Nona called him 'Teddy Bear' and introduced me as her twice-removed uncle from London. We began talking to each other and soon discovered that we had both been in the Battle of Arnhem. 'Teddy' was Major Winrich Behr, third staff officer under General Krebs, and had been based in the German headquarters in the Hotel Hartenstein at Oosterbeek. Within hours of the British landing, these headquarters became ours.

When we talked I became more and more fascinated as I learnt about the German view of the battle. Then – over forty-five years ago – I had, of course, no idea of what the Allied overall strategy was, nor of the reasons for our defeat at Arnhem. I asked Teddy to tell me more about himself and everything he could remember about the battle. We didn't want to bore the other guests, so we agreed to continue our discussions at another meeting, and to write to each other.

I learnt that he had been born in Berlin on 22 January, 1918, and had been christened Winrich Hans Hubertus Behr. He came from a long line of Prussian Army officers and civil servants. His father, sword in hand, had led his regiment into the first great battle of the First World War at Maubeuge. He had been seriously wounded and had retired with the rank of colonel. Winrich had been brought up by his well-to-do family in the Tiergarten district of Berlin. When the Nazis came to power in 1933 he was a schoolboy of 15 and became fired by the new political ideas and the National Socialist propaganda. They had promised to ensure that Germany would break free from the disgrace of the Versailles Peace Treaty; they would put an end to unemployment and political corruption, and would establish a firm regime based on law and order. The Hitler Youth, to which most of Winrich's school friends belonged, with its smart uniforms, its feeling of togetherness and belonging, and its adventurous out-of-door activities, strongly appealed to him. His father, however, categorically forbade him to join. Colonel Behr was a staunch conservative and had supported Hugenberg's German National Party (Deutsche National Partei) which, although it had helped to get the National Socialists into power, was banned by them as soon as Hitler became chancellor. Winrich's father was a staunch anti-Nazi and totally opposed to 'that bricklayers' apprentice, Adolf Hitler'. But in 1935 the entire youth section of the tennis club, in which Winrich was one of the top players, was forcibly incorporated into the Hitler Youth and he found himself a member of the Nazi cadet party in spite of his father's objections. The same year he passed his final examinations at the French Grammar School. This was followed by six months' compulsory service in the Arbeitsdienst ('Labour Service' – a work battalion), before he joined an armoured regiment near Potsdam for his basic military training. He was soon transferred to the Military School in Munich, where he was commissioned in January 1938.

In the spring of that year he took part in the occupation of Czechoslovakia and in the autumn of 1939 in the invasion of Poland. Only then did he realise that Hitler might lead Germany into a full-scale European war. He hoped that such a disaster might be averted, however – perhaps through an understanding with the British, who had always seemed to approve of Hitler's European military adventures, and had even been a party to the dismemberment of Czechoslovakia.

But Britain and France declared war. Poland was conquered very quickly, danger from the east having been averted in August when Hitler signed a very opportunistic treaty with Stalin. Russia was now no threat and Behr became one of the thousands of troops transferred to the Western Front in 1940. He was appointed commander of a tank company with the rank of captain and was part of the triumphant drive through the Netherlands, Belgium and France which ended with the French agreeing to an armistice and the remnants of the British Army escaping from Dunkirk.

In the bitter winter of 1941 Behr's Panzer Reconnaissance Unit was sent ahead by train to Naples and then to the welcome sunshine of Tripoli in North Africa. As soon as they landed they were despatched at full speed to El Agheila to join Rommel's Afrika Korps. They arrived in time to take part in the triumphant advance through Tobruk and Benghazi and on via Sollum and Halfaya Pass, well beyond the frontiers of Egypt. The company was employed as a reconnaissance unit, probing ahead of the main army in fast light tanks and reporting back the positions and strengths of the British armour. Often Behr was reporting directly to Rommel. He was very impressed by the field marshal's leadership and his military skill; he also found him a warm and understanding human being.

Behr's operations were so successful that Rommel decorated him with the *Ritterkreuz*, the Knight's Cross, Germany's highest military decoration; he had already been awarded the Iron Cross Second Class in Poland in 1939, and the Iron Cross First Class in France in 1940. At this time in the desert Behr could still believe in a final German victory. Their tanks and armament were superior to those of the British and for a time they even had air superiority. Both sides fought honourably and well and scrupulously observed the Geneva Convention in the treatment of POWs. Behr spent some time talking to British prisoners, finding that they, too, believed in a final victory, although there was general agreement that war was a horrible and futile activity.

By August 1942, however, the tide had clearly turned for the Afrika Korps and they began a slow retreat westwards towards Tobruk. The British had been receiving substantial reinforcements of men and equipment and, even more importantly, had a new and brilliant Commander-in-Chief: Montgomery.

Behr did not have to experience the ultimate humiliating defeat of the Afrika Korps and its expulsion from North Africa. In his last reconnaissance his unit was surprised by a squadron of the new American Sherman tanks. They managed to get away, but not without taking several direct hits. With shrapnel in his chest and upper arms, Behr was flown out to Germany and a military hospital. His convalescence was followed by an undemanding job as an instructor in a military school near Potsdam. This quiet life lasted only a short time and then he was assigned to the Sixth Army, commanded by General von Paulus, at that moment fighting the Russian Army on the Eastern Front. In October 1942, he joined the staff of the Sixth Army headquarters in Golubinka on the River Don, about 30 miles west of Stalingrad. There Behr found himself in charge of artillery, as well as being responsible for keeping up-to-date maps of the military situation and for dealing with all incoming and outgoing reports on the progress of the battle that was raging at Stalingrad.

The great Russian attack began on 20 November. Behr and Oberst Elchlepp, the first general staff officer, flew the next day in a Fieseler Storch reconnaissance aircraft to Gumrak, the nearest landing strip to Stalingrad. From there they made their way on foot through the thunder and smoke of the barrage until they reached General Paulus's bunker, where they helped to organise the communications. Behr reported daily to Paulus on the general situation in Stalingrad – a situation that became more desperate each day.

With the onset of the Russian winter, hunger, disease and the bitter cold began to take a greater toll than the relentless bombardment from the encircling Russian armies. And still the Führer's orders were to hold on to their positions and then counter-attack. They knew the situation was hopeless, but Hitler and his General Staff would take no notice of their desperate pleas either for reinforcements or for permission to surrender. On 13 January 1943, Paulus ordered Behr to fly out to report in person to Hitler in the Wolfsschanze – the 'Wolf's Lair'. Paulus said that Hitler had lost all faith in his generals: he hoped that he might listen instead to an 'ordinary', though highly decorated, tank officer.

Behr's plane took off in the late afternoon, just before the airfield fell into Russian hands. He landed in Taganrog near the Black Sea, reported to Field Marshal von Manstein and, on the morning of the 14th, flew to East Prussia. That evening he reported to Hitler's headquarters, and there met the Führer and his General Staff for three and a half hours. All the top generals were there – Field Marshal Keitel, commanding the Wehrmacht, Colonel-General Jodl, Head of Operations, Major-General Schmundt, Hitler's chief adjutant, and Martin Bormann, Hitler's deputy.

Captain Behr presented an extremely clear picture of the situation facing the encircled armies in Russia. The Germans were under continuous bombardment; the cold was intense – some 30 degrees Celsius below zero; there was no shelter and no transport – even the pack horses had been eaten, Behr wanted to know what help they could expect. This was the first time Hitler had heard the truth about Stalingrad. He seemed stunned, then responded with a rapid outline of his plans for an offensive to start in six weeks. He complained bitterly about all the mistakes that had been made and detailed the situation in the various sectors, but he did not breathe a word about support, or supplies, or reinforcements. Behr was unable to remain silent. In spite of himself, he interrupted Hitler. The army, he said, was at the end of its strength: it was essential that he could tell them what supplies they could expect within the next forty-eight hours: that was all the time they had left. No one had ever interrupted the Führer before. There was dead silence. Everyone expected an explosion, but none came. Hitler remained calm, and quietly ordered Field Marshal Milch of the Luftwaffe to look into the possibility of supplying help from the air. The conference was over. And Behr was now convinced that the war was lost.

Behr was lucky. He had to stay at Hitler's headquarters because there were now no more flights to the Eastern Front. But the thought of the helpless suffering of his comrades in Stalingrad hung over him like a dead weight and he was unable to rejoice in his own escape. In the comparative peace and luxury of the Wolfsschanze he followed the disastrous news from Stalingrad hour by hour. He wondered how much longer they could hold out. They lasted another three weeks before Field Marshal Paulus (he was promoted in one of Hitler's last orders to the beleaguered garrison) surrendered with seven other generals and 45,000 officers and men. In the whole Stalingrad operation 200,000 Germans were killed and some 100,000 taken prisoner, of whom fewer than 4,000 managed to get back to Germany after the war.

After Stalingrad, Behr was posted to command an infantry battalion in the Caucasus, where he served for seven months; then, promoted to the rank of major in the General Staff – with an impressive red stripe down his trousers – he was attached as staff officer to Army Group 'B' in northern France, first at Fontainebleau and then at La Roche Guyon on the Seine. Here he was pleased to be reporting once again directly to Rommel. But he found the field marshal greatly changed; he was in poor health, and lacked his former drive and warmth of feeling. (Later, in suspicious circumstances, Rommel committed suicide, having been badly wounded in France in July 1944.) Behr was then responsible successively to Field Marshal von Kluge, General Krebs, and, finally, Field Marshal Model. It was as first assistant staff officer to General Krebs that Behr took part in the Battle of Arnhem, an impression of which the German point of view is given in the next chapter. He remained with Model until the very end, trying to save as many men and as much material as possible during the retreat of German armies in north-west Europe, following the Allied invasion of France in June 1944.

In December 1944, Hitler ordered a massive offensive, under the overall command of Field Marshal von Rundstedt, through the Ardennes Forest in Belgium which, he claimed, would finally stop the retreat by breaking the Allied front in the west and cutting their supply lines. Model still believed in the Führer and in the possibility of a German victory, and threw himself enthusiastically into the planning of this daring project. Behr had no such faith, but gave Model his full cooperation and stayed by his side throughout the ensuing 'Battle of the Bulge'. When, by the middle of January 1945, it was clear that the battle was lost, Model committed suicide and Major Behr changed into civilian clothes and went home to join his family.

As soon as he could after the end of the war, Winrich Behr went to Bonn University and then to Frankfurt, where he studied law. He qualified as a lawyer in Düsseldorf in 1952, and was then recruited by Jean Monnet, the founder of the European Coal and Fuel Community – the forerunner of the European Common Market – and in time became head of the private office of the deputy president. He was called to Brussels by his former teacher, Professor Doctor Walther Hallstein, and there became deputy general secretary of the European Economic Community (EEC). In 1965 he was appointed managing director of Telefonbau and Normalzeit, a business manufacturing private telephones and switchboards. Then it had a turnover of 400 million marks and employed 12,000 people. When Winrich Behr retired in 1983 the turnover had increased to 2.3 billion marks.

At the time of writing, in his lovely country house near Düsseldorf he is very busy – he is on the board of several industrial concerns and on committees concerned with Prussian country houses and gardens, and European cultural institutions. He describes himself as a European – but he is actually a very admirable citizen of the world.

5

The German View of the Battle

Sunday

On this sunny Sunday morning, while Louis Hagen was sitting in the NAAFI canteen on Windrush airfield in East Anglia, waiting to hear when his glider section would be despatched to the Netherlands, Major Winrich Behr, first assistant staff officer to General Krebs, the Chief of Staff, had just finished his lunch in the Hotel Hartenstein in the suburb of Oosterbeek, 2 miles west of the town of Arnhem. It was a beautiful day and he was enjoying an unaccustomed and very welcome feeling of relaxation.

For more than three months, since the Allied invasion of Normandy on 6 June 1944, the Germans had been steadily pushed back across Europe by an enemy with greatly superior armour and firepower. Their slow and bloody retreat out of France, through Belgium and into the Netherlands, had at last come to a halt. It seemed that their dogged and relentless pursuers had finally run out of steam. They now had three major rivers between them and the enemy; Behr felt it was about time for a quiet breathing space.

He had had a long and exciting war: he had won his *Ritterkreuz* with Rommel in North Africa and had been one of the last officers to escape from the frozen shambles of Stalingrad. His present situation was almost routine compared with his previous experiences.

But there was much to do. All round Arnhem the Germans had set up a series of field workshops and transit camps where the stragglers and survivors of the long retreat could be collected together. Many divisions had been reduced to a fraction of their original strength and were now being regrouped into new operational units. Every day specialised freight trains brought back the battered tanks and mobile guns which had stubbornly held back the Allied advance, allowing the infantry and other troops to fall back towards Germany. Here in relative peace the fitters and engineers worked on urgent repairs and refitting; the weapons were repaired or replaced and the crews re-equipped and retrained. Then, as soon as the tanks, SP guns and fighting vehicles were ready for action, they were sent eastwards without delay to help in the defence of Germany.

There was an enormous concentration of heavy armour in all stages of preparation, from cannibalised wrecks to fully battle-ready Tigers. Some of these were the updated Royal Tigers, with much thicker armour plating and larger guns, which had proved a match for the Russian T34s. Among its armament was the 88mm gun (originally an anti-aircraft weapon) that had wrought such havoc among the British tanks in North Africa. The Hotel Tafelberg, also in Oosterbeek and close to the Hartenstein, was the headquarters of Generalfeldmarschall Walter Model; commanding Army Group 'B'. Winrich Behr's immediate boss, Krebs, was his Chief of Staff and was based in the Hotel Feldberg nearby. Model, a typical heavily built professional soldier in the Prussian tradition – he even wore a monocle – had succeeded Generalfeldmarschall Gerd von Rundstedt on 5 September in command of all the German armies between the Dutch coast and Alsace-Lorraine.

Major Behr's quiet lunch was suddenly shattered by the roar of a squadron of fighter-bombers sweeping over the hotel, guns firing. There was a shocked silence and then a babble of excited discussion; the officers present quickly realised that the aircraft had not been firing at them; although in the next room the staff artillery officer, Oberleutnant von Metzsch, had had his soup spoon shot out of his hand.

There had been no preliminary warning of any expected offensive, but the considerable number of aircraft involved strongly indicated that something unpleasant was brewing. Major Behr was suddenly aware of extraordinary goings-on outside the window. He couldn't really believe it, but there, not much more than 300 yards from the hotel, a group of parachutists was silently floating down.

While everyone scrambled to get their weapons, Behr sent a couple of soldiers up to the roof to report what they could see. His first thought was that this was a commando raid aimed at capturing some or all of the senior officers. As well as the Generalfeldmarschall and Krebs, there were at least three other senior generals in the neighbourhood. Behr hurried off to the Hotel Feldberg to report to Krebs.

Some 3 miles to the north-west, in the Wolfheze Hotel, Sturmbannführer Hans Kraft watched appalled as a mass of gliders silently swept out of the sky. His SS Panzer-Grenadier Training and Depot Battalion, being composed largely of reluctant teenagers and elderly conscripts, were hardly crack front-line troops, but he had no time to worry about that. He rushed out into the courtyard and yelled for all ranks, including trainees and wounded, to grab their weapons and ammunition and take up action stations; the men nervously scurried to and fro. Kraft telephoned his superior, Obergruppenführer Wilhelm Bittrich, commanding II SS Panzer Corps, who told him that there were also reports of airborne troops dropping near Oosterbeek. He ordered Kraft to turn his headquarters into a defended strong-point and to send out armed patrols to find out what the enemy were doing. Reinforcements would be sent as soon as possible.

When Major Behr reached the Hotel Tafelberg he found that Krebs had already received reports of glider landings and that Model was hurriedly packing and had ordered the entire General Staff to move out to Bittrich's headquarters in Doetinchem, some 20 miles to the east.

General Krebs told Behr to find out if there were any enemy troops to the north or east of Arnhem and to prepare transport. Behr rushed back to the Hartenstein, where the men from the roof reported that they had seen a mass of planes, troop- and cargo-carrying gliders, and parachutists, mostly to the west and north of the town.

Somewhat to Behr's surprise, all the telephone lines were open and the roads back to the east were clear. Immediate preparations were made to evacuate. By this time they were so used to the routine of sudden departures that there was no undue excitement. During the past six months they had had to move their headquarters so many times that there were now comparatively few secret documents or codes to burn. It was really no more than a matter of packing up shaving kit and spare clothes and piling into the waiting transport.

Within half an hour Behr was in his BMW staff car, driving away from Arnhem as part of a convoy headed by Model in his bullet-proof personnel carrier, with an escort of four light tanks.

Back at Wolfheze, Sturmbannführer Kraft's patrols were reporting that several hundred enemy soldiers had landed and were being marshalled into columns, apparently preparing to move into the town. Pleased by the response of his untried young troops, Kraft sent a larger group with mortars and anti-tank guns to attack the troops in the landing area. Other groups he told to get ahead of the invaders on the road to Oosterbeek and set up ambushes and strong-points to impede them as much as possible.

Doetinchem was where Obergruppenführer Bittrich had set up his corps' logistic headquarters. He was a good-looking regular officer who had joined the Waffen SS as a tactical career move, without having any kind of commitment to National Socialism. Bittrich was the overall area commander; his troops included the surviving units of the 9th SS (Hohenstaufen) Panzer Division and the 10th SS (Frundsberg) Panzer Division. He had been warned of the imminent arrival of the Generalfeldmarschall and his staff and gave instructions for accommodation to be found in houses and farms nearby.

Bittrich now began to prepare the counter-attack. He ordered Obersturmbannführer Walter Harzer, 9th SS Panzer's Chief of Staff and temporary commander, to send a reconnaissance group in some strength to check out the country between Arnhem and Nijmegen to the south, and to secure the bridge over the Waal in the latter town. He wanted another force to occupy Arnhem, reinforce the guard at the bridge and hold it until Brigadeführer Heinz Harmel's 10th SS Panzer could take over the responsibility. Once the bridge was secure and handed over to the men from the Frundsberg Division, the Hohenstaufen Panzers were to destroy the enemy forces near Oosterbeek.

The 9th SS Panzer Division had been refitting for the past three or four weeks in their staging-post near Apeldoorn. Recently Harzer had been told to transfer several of his tanks and armoured cars to Harmel's Frundsberg Division, which was in a much weaker state, and which had been ordered back to Berlin. Harzer had decided to keep back some of his tanks, however, and the workshops had accordingly been told to remove tracks or wheels from those in best condition and then report them unserviceable. Now, following the Obergruppenführer's orders, the tanks had quickly to be put back into fighting order. But on this lovely September Sunday most of the men were off duty. Twenty trucks were sent to quarter the town picking up any 9th Panzer men they saw and to tell all the others to pass on the word for all ranks to join in the defence of the town, making their way to wherever they heard shooting. Although the engineers and fitters worked with record speed, it was not till late afternoon that a sufficient number of tanks were battle-ready.

Harzer had now divided his division into two groups (Kampfgruppen), and given Sturmbannführer Brinkmann responsibility for checking the road to Nijmegen while Sturmbannführer Spindler was to hold the bridge, occupy Arnhem and deal with the forces at Oosterbeek. It was late afternoon before Spindler's group drove into Arnhem and reported that the town was deserted.

Shortly afterwards Brinkmann's forty-vehicle reconnaissance force drove over the bridge towards Nijmegen.

As all was quiet at the bridge, Spindler decided to drive on to where the action was. Beyond the St Elizabeth Hospital there was a lot of shelling and machine-gun fire: the invading force was being held up by Kraft's battling teenagers from the training battalion, reinforced by stray soldiers alerted by the 9th SS Panzer trucks when they were rounding up their troops. Spindler left four tanks to reinforce the defenders and with the rest of his group he set up barricades across the two main roads leading from Oosterbeek into Arnhem. Harzer also sent another group from the Hohenstaufen Panzers to guard the roads from Ede and Utrecht. Model, meanwhile, had set up his new headquarters in a little castle in the town of Terborg, about 5 miles away from Doetinchem. Reports from Arnhem were continually arriving there, and Model called a staff meeting to evaluate the situation. It was soon clear that there was no danger of any immediate breakthrough in strength. All the reports coming in showed that there were more than enough troops and armour in and around Arnhem to deal with the enemy threat – whatever their intentions might be.

So what was the enemy up to? The possibility of a raid on his headquarters had very quickly been discounted: it had been perfect dropping weather with a very light breeze, so they had to assume that the obviously experienced parachutists were dropping where they had meant to drop. In which case, the main force had landed too far away from the General Headquarters to mount a surprise attack.

For the same reasons, Model insisted that the road bridge could not have been the target. A complete brigade, dropped as close to the southern end of the bridge as the soft ground allowed, could have overcome the small guard posts and held the bridge against determined opposition for a significant length of time. But from the reports it was clear that all the fighting had been on the outskirts of the town and that no troops had dropped anywhere near the road or rail bridges.

What did that leave? An attempt to destroy the regrouping Panzer brigades? This was one of the earlier theories, and a number of nearly battle-ready Tiger tanks had quickly been loaded on to two trains, ready to pull out to the east. But as no sign of the enemy's interest in the tank repair workshops had developed, the trains were unloaded and the Tigers returned to their workshops.

The headquarters staff assumed that the enemy would have had complete intelligence about the concentration of armour round Arnhem. All the activities connected with moving and repairing heavy tanks could not have been hidden from the local population, some of whom were undoubtedly in touch with the Dutch underground. And however carefully they camouflaged their workshops, the enemy's air reconnaissance must have revealed the tell-tale signs of continuous tank movements. But again, the force that had landed was much too small to engage and destroy the German armour; a squadron of heavy bombers would have been much more appropriate. The only possible explanation was that the landings were a diversion intended to keep a large number of troops and armour engaged with the airborne forces while a surprise ground attack took place somewhere else.

At Model's HQ Winrich Behr could not spend too much time on speculation. The myriad administrative details which flowed from their sudden move would keep him busy for several days. In a cramped little office at the top of the castle he reached for the telephone to start linking up with the local authorities and services, informing everyone who might need to know where he and General Krebs were now operating.

Bittrich tracked down the Generalfeldmarschall in his new headquarters in Terborg and outlined his plans, which Model immediately approved. Meanwhile reports had come in of heavy fighting against American airborne troops near Nijmegen and Eindhoven, and Model now said that he was sure that the ultimate objective of the drop was the Ruhr. All their energies should be devoted to blocking any moves by the enemy to break out in a north-easterly direction. Bittrich argued that, in that case, surely they should now blow the bridges at Arnhem and Nijmegen. Model absolutely refused, however, saying that the bridges were essential for the deployment of German armour.

Back in Arnhem, there was intermittent firing all night as Spindler's force attacked pockets of enemy troops along both sides of the railway, and at the north end of the bridge.

Monday

Early in the morning the reconnaissance group from Harzer's 9th SS Panzer returned from Nijmegen where they had had a sharp engagement with an American force holding one of the two bridges over the Waal. They had driven them off with few losses and were returning to join up with the rest of the division. They drove unsuspectingly straight over Arnhem Bridge, only to be met with enemy small-arms and anti-tank fire: overnight the north end of the bridge had been captured by the British airborne troops. Four scout cars had raced across and into the town before they were blown up. Eight half-track vehicles following were destroyed on the bridge and the rest of the group quickly reversed to the shelter of the south bank.

Sturmbannführer Kraft had been warned to expect another landing further west of Wolfheze. When his patrols reported enemy troops defending the Groote Heide, an area of heathland where the most westerly of the British DZs were, he sent out three groups with machine guns to deal with them. There followed a continuous exchange of fire with casualties on both sides, enlivened at one point by a Luftwaffe squadron dropping a carpet of bombs on the open space between the combatants. And shortly after that came an even more worrying event – the arrival of the anticipated landing. Suddenly the sky was filled with hundreds of gliders and parachutists.[6] To Kraft's teenagers from the Panzer Training Brigade, the sky seemed to be blotted out by a thousand parachutes. They knew that this was no place for them and hurried back to Wolfheze.

Obergruppenführer Bittrich was alarmed when he heard Kraft's report on the size of the enemy landings. He felt very much under pressure. He was still waiting for the promised reinforcements under Generalleutnant Hans von Tettau, whose scratch 'division' had been stationed on the River Waal to collect stragglers from the fighting on the Maas and re-form them into fighting units, and was now to the west of Arnhem. Bittrich now ordered Kraft to send his entire force to join up with Spindler's group from 9th SS Panzer, which was already fighting in the town itself. There the north end of the road bridge was held by

---

6 The second pilot in one of the gliders approaching the landing zone at Wolfheze was Louis Hagen.

the enemy,[7] due to the inexplicable negligence of Harzer who had been told specifically to hold the bridge until relieved by 10th SS Panzer. Now it was going to be very difficult to get the tanks from 10th Panzer over to the south bank of the river where they would be needed to deal with any further enemy landings in the Nijmegen area. Bittrich therefore told Harmel that he would now have to ferry his division across the river at Pannerden, 7½ miles south-east of Arnhem. But it would be a slow business.

Tuesday

Now that the scale of the enemy operations seemed to be clearer, Model moved his headquarters back to just north of Arnhem, where Harzer had his command post in the Heseburgherweg School. By now von Tettau had arrived with an ad hoc fighting force of administrative and training units, an SS depot battalion, an NCOs' School and a battalion of Dutch SS men, and was ordered by Bittrich to guard the area of the landing zones to the north-west of Oosterbeek. Von Tettau found these in the control of the Allied airborne troops and sent Standartenführer Lippert (the commanding officer of the NCOs' School) with six old Renault tanks and a company of infantry to clear them out. The tanks with their untrained crews were swiftly dealt with by British Piat anti-tank weapons. But, following a feint attack by Lippert, the scratch German infantry charged the airborne defenders, put them to flight and pursued them towards Arnhem.

When Lippert returned to the landing zone he found that a large parachute and glider landing had taken place and had great difficulty getting back to von Tettau's headquarters, although he managed to round up a few dozen British prisoners on the way.

---

7 This was a force from 1 Parachute Brigade under Lieutenant-Colonel John Frost, mainly Frost's own 2nd Battalion Parachute Regiment. As they had advanced from their landing zone into Arnhem, the railway bridge was demolished just as they reached it, and they found the pontoon bridge partially dismantled. The road bridge was thus the only practicable crossing for miles.

Wednesday

At dawn the heavy SS Frundsberg Mortar Section moved into position and blasted the Arnhem bridgehead. This was followed by a frontal attack by ten somewhat elderly tanks firing wildly but continuously, and supported by infantry keeping up a steady pounding of heavy machine-gun fire. The tanks were met by the very accurate fire of the British 6-pounder anti-tank guns; they slowed, came to a stop and then began to back away. The barrage quietened and the machine-gunners slipped back with the tanks. But, from a safer distance, the SS mortars kept up a continuous fire.

Reinforcements continued to pour into the battle area. Model could now report to Hitler that the larger part of the Fifteenth Army in the Antwerp area, which Montgomery's forces had by-passed during their advance, leaving it to escape largely intact, had now crossed the River Scheldt and were in a position to move against the enemy near Nijmegen. Here the narrow corridor of enemy transport was also being attacked by II Parachute Corps from the Reichswald Forest.

There was an immediate and unexpected gain from the fighting round Nijmegen. Allied battle plans had been found on the body of an American officer whose glider had crashed near the town. They included details of an imminent new landing at the village of Johannahoeve, west of Arnhem but east of Wolfheze. Model ordered Kraft's SS Panzer-Grenadier Training and Reserve Battalion to surround the landing zone.

They found enemy troops already there and immediately engaged them. The battle was at its fiercest when the landings started; the luckless glider and parachute troops dropped in the midst of the fighting and many were killed as they landed.

Thursday

It was a time of almost total confusion with no clear line of separation between the two forces. It seemed as though there were three distinct battles. Elements of 10th SS Panzer were attacking the British forces holding the north end of the Arnhem road bridge, Kraft's training battalion, with reinforcements from 9th SS Panzer, were doing their best to block three separate thrusts from the west by airborne troops desperate to relieve the forces holding the bridge. Here, between the Rhine and the centre of Arnhem, British troops were doggedly fighting their way towards the bridge street by street, and were suffering considerably in the process.

The invading force's headquarters was now in the Hotel Hartenstein in Oosterbeek, and the third battle involved 9th SS Panzer, which had formed a ring of armour round this headquarters perimeter and were subjecting the enemy to a continuous artillery barrage. The fighting in the western suburbs and the Oosterbeek area was even closer and more concentrated than elsewhere, with the Germans sometimes occupying houses next door to those held by the enemy.[8] In some cases the opposing sides even found themselves on different floors of the same house.

Model was annoyed and worried: he could not understand how it was possible for the tiny British force still to be holding the Arnhem bridgehead, thus preventing the German heavy armour moving south where they would be needed to start clearing the area around Nijmegen. He told Bittrich in uncompromising terms to get something done about it. Harzer was to use whatever forces were needed and free the bridgehead within twenty-four hours.

To help the operation Model ordered the evacuation of all civilians from Arnhem and told the Luftwaffe to mount a heavy bombing raid on the town centre. Convoys of lorries and ambulances took out the women and children. 9th SS Panzer were made responsible for the collection and assembly of the many enemy prisoners, and also the wounded of both sides.

The bombers pounded the already wrecked city centre, making house-to-house fighting more difficult for both sides. By evening the

---

8 In one of these Louis Hagen was fighting from an attic window with a Piat and a Bren gun.

Dutch police, firemen, air-raid wardens, doctors and nurses were the only civilians left in Arnhem.

Meanwhile the fighting round the bridgehead had reached a new intensity. Many of the defended buildings were on fire and the British wounded were in danger of being burnt alive. Under the protection of white flags, officers of 9th SS Panzer agreed with the British officer in charge, Major Freddie Gough[9] – whose senior, Colonel Frost, had been wounded – to a temporary ceasefire while the casualties were rescued from the burning houses.

But the battle of the bridgehead was nearly at an end. That night the intensive pounding of the British positions at the bridge by the Kampfgruppe Brinkmann artillery reached a climax. There was very little response coming from the trapped British troops.

Friday

On the Friday morning things were looking a lot better to Obergruppenführer Bittrich. The first news was that the bridgehead had been finally captured and that Major Knaust with eight Panther tanks had swept aside the wrecks of the vehicles that had been destroyed on the bridge and was racing south to join the fighting at Nijmegen. Most of the remaining enemy forces were now corralled in a small enclave round the Hotel Hartenstein in Oosterbeek, and German reinforcements had been flowing in from all points of the compass – a motley collection of units ranging from battle-hardened tank men to 16-year-old 'non-combatants'. Hauptsturmführer von Allworden reported with a unit from the SS Panzer Pursuit Corps, and Obersturmbannführer Harder with an SS Panzer Company. And Bittrich had another stroke of luck. Full details of the proposed landing

---

9   Gough in fact commanded 1st Airborne Division's Reconnaissance Squadron, some elements of which had reached the bridge.

zones for a new supply drop had been found in a crashed British glider. So when Harzer got news of a further airborne force crossing the coast at Dunkirk – a large force with many heavy transport aircraft travelling south-east, clearly making for Arnhem – Bittrich was ready. He ordered a general alert and warned Spindler to increase the number of troops guarding the proposed landing zones, which the British had already marked with large coloured rectangles for different kinds of stores, as detailed in the captured plans. His task was made easier because the 503rd Panzer Division had just arrived with an additional forty-five Royal Tiger tanks. Also a Panzer-Grenadier battalion from Berlin was unloading at the rail terminal.

At the landing zones anti-aircraft guns were prepared for action, everyone waiting for the expected arrival time – 1600 hours. The tension grew; then the Luftwaffe gave the first indication of the arrival of the airborne force. A squadron of German fighters roared over the town, immediately splitting up as escorting enemy fighters rushed in to head them off. The anti-aircraft guns pounded the skies as the black mass of planes neared the town. Small-arms and machine-gun fire joined in. The slowly descending transport planes and parachutes dropped helplessly into a waiting inferno.

The German troops surrounding the landing zone soon realised that most of the parachutes were marked with red and white stripes. They knew that this meant that they were bearing supplies, not armed soldiers. Gradually the firing died down and the men watched as the Allied planes braved the barrage of flak to drop a rich harvest directly into their enemies' hands. The Germans gained a much appreciated haul of medical supplies, emergency rations, bacon, milk powder, bread, flour and even chocolate. This 'Father Christmas' drop cost the enemy dearly in crashed planes and dead aircrew.

By now the condition of the wounded – both British and German – within the Hartstein compound was desperate. There was a shortage of drinking water, as well as medical supplies. As a humanitarian gesture Harzer, commander of the 9th SS Panzer Division, sent

Hauptsturmführer Doktor Skalka to negotiate a ceasefire and to offer to have all the wounded moved out to hospitals in the town.

Dr Skalka drove into the compound with a British soldier sitting on the bonnet of his car holding a large Red Cross flag. He was met by 1st Airborne Division senior MO, Colonel Graeme Warrack, who got Urquhart's permission to accompany Skalka back to 9th SS Panzer headquarters. Arrangements were quickly completed, and as a result a ceasefire was accepted by both sides. A convoy of German ambulances drove into the Allied compound to bring out all the seriously wounded German troops, as well as many British and Polish soldiers. They were distributed among field hospitals in Arnhem, and in the St Elizabeth Hospital which was still staffed by British doctors and orderlies under German supervision.

Saturday

Bittrich now ordered all German forces to concentrate on the pocket of enemy troops centred round Oosterbeek.

At the same time forty-five Tiger tanks of 503rd Panzer Division, under von Tettau, with supporting infantry, began attacking the airborne troops who were dug in round the Heveadorp ferry. Many Allied troops were killed or wounded, but by the late afternoon 503rd Panzer had failed to gain any ground.

Harzer's Hohenstaufen tanks were holding the area to the north and Kraft was attacking from Oosterbeek on the east. By the afternoon several streets had been cleared, in spite of desperate defence by the enemy using pistols and hand grenades. There were heavy losses on both sides. Several of the German commanders commented on the incredible bravery of the Allied troops, by now heavily outnumbered and completely outgunned. At dusk the Frundsberg and Hohenstaufen tanks probed into the enemy pocket, enabling snipers to take up positions in trees and empty buildings.

Sunday

Next day tanks from 9th SS Panzer attacked along the northern rim of the Hartenstein perimeter, but were driven off by artillery fire from the south bank of the Rhine – elements of Second Army had at last reached Arnhem, though too late to do much for 1st Airborne. The Germans nicknamed the concentrated pocket of resistance 'Der Hexenkessel' – the witches' cauldron. By now a continuous pall of black smoke hung over the cauldron where the Allied airborne troops were fighting to survive. The acrid stench of high-explosive mixed with the oily reek of burning vehicles. Intermittent flashes of artillery fire and exploding mortar shells only emphasised the surrounding gloom.

Frustrated by the prolonged stalemate, Harzer decided to try a psychological approach. He had loudspeakers set up in the trees round the perimeter; through them, the enemy were told how hopeless their situation was, that no reinforcements could possibly reach them, and that good treatment, food and drink awaited them if they only gave up their useless resistance. These attempts were met with defiant and scornful shouts from the defending airborne troops.

Monday

Next day Bittrich ordered attacks to be redoubled on all sides of the 'cauldron'. He had been pleased to learn that fifteen Royal Tiger tanks were on their way to reinforce 9th SS Panzer on the east of the perimeter. Together with flamethrowers from the 9th Pioneer Instructing Battalion, they were now able to overcome individual strong-points in the cauldron. But it was slow work and the Royal Tigers, massive and virtually impregnable, proved unwieldy and awkward in the narrow and rubble-strewn streets of Oosterbeek.

On the eastern side Harzer regrouped the SS Hohenstaufen into four separate spearhead assault groups under Sturmbannführers

Mollers, Spindler, von Allworden and Harder. In the afternoon von Allworden and Harder broke through and destroyed two British artillery positions, but were halted from advancing further by artillery fire from elsewhere within the perimeter. Bittrich was now getting reports that the enemy fire seemed to be getting weaker and more sporadic. He passed the good news on to Model, who reported in turn to von Rundstedt (the Commander-in-Chief West, and thus Model's immediate superior), telling him that in order to destroy the extremely stubborn enemy he had to have reinforcements, including at least a Panzer brigade with infantry and an artillery battalion.

Meanwhile, more artillery was being brought into action. All the available heavy guns of the 191st Artillery Regiment were targeted on Oosterbeek and then began a continuous and head-splitting barrage over the whole of the seething cauldron. Now, adding to the excruciating din, the Allied artillery on the south bank of the Rhine began a continuous attack on 191st Artillery Regiment's gun positions. This exceptional concentration of fire from the enemy heavy artillery was a worrying development.

Tuesday

Shortly after midnight men of Kampfgruppe Harder reported that several boats had tried to cross from the south bank, but had been driven back or sunk by mortar and rifle fire. Bittrich began to wonder whether the enemy might now be in a position to send in reinforcements, and so turn the troops in the Hartenstein pocket once more into a dangerous attacking force. He ordered Harder to keep a steady mortar barrage over the river and on to the southern bank for the rest of the night.[10]

At first light Bittrich ordered redoubled attacks on all sides of the smoking and tangled wreckage and ruins that were all that was left of the cauldron. The Hohenstaufen tanks pushed forward, surprised at

---

10   At about this time Louis Hagen was swimming across the Rhine to safety.

the thinness of the defending fire. Then, eerily, the firing from within the perimeter seemed to peter out. The sudden silence was unsettling. Gingerly the tanks from the two Kampfgruppen edged their way towards the Hotel Hartenstein, infantry fanning out behind them. The tanks reached the hotel and the infantry circled round it silently. They reached the tennis courts and were suddenly greeted with a chorus of German voices. This was the prisoners' compound, and the Hohenstaufen men now learnt that it really was all over. The prisoners told them that the Allied troops had pulled out overnight; all that were left were sick and wounded men who had been firing at random while the main body of defenders escaped across the river.

It was over. The German soldiers couldn't believe it. During all the turmoil of the night's artillery and mortar duel, the apparently invincible airborne fighters – or at least the survivors – had slipped unnoticed across the Lower Rhine.

Major Winrich Behr learnt with amazement that the Allied forces that launched the airborne attack, of which Generalfeldmarschall Model had made so light, had fought on for nine days against, among other troops, two elite SS armoured divisions. It had been quite a problem after all, but a problem that had eventually brought its own rewards. The Germans had made enormous gains. Many British officers and thousands of parachute and airborne soldiers had been taken prisoner. In the Arnhem–Oosterbeek area, the 9th SS Panzer Brigade had captured some 6,000 British and Polish prisoners. The booty had included an arsenal of weapons, ammunition and equipment, including a number of light tanks and anti-tank guns, innumerable jeeps and field howitzers, 250 field weapons of all kinds, 1,000 gliders, untold numbers of rifles and other small arms, and an enormous quantity of supplies. Nearly a hundred RAF aircraft had been destroyed. And in the course of the nine-day struggle, an elite and highly trained enemy fighting division had been destroyed.

6

Summing Up

When I started to write the background to this book, nearly fifty years after its first publication, I read most major accounts of the Battle of Arnhem in both the English and the German archives. Gradually I began to understand the overall strategy – and the muddle, the heroism and the waste. Quite apart from the unpredictability of any battle, as well as several strategic errors, I have come to believe that there was one major reason for the British defeat at Arnhem – the overwhelming ambition of Field Marshal Montgomery. He was a brilliant leader, full of imaginative ideas, but he was an emotionally withdrawn man. He had a reputation as a meticulous strategist who would never move without overwhelming odds in his favour. This time his judgement failed. He resented being subordinate to General Eisenhower and was frustrated by what he saw as the Americans and Russians. His arrogant ambition discounted adverse intelligence reports prior to 'Market Garden', and doubts expressed by senior staff officers.

At first Eisenhower had opposed the plan, but Montgomery's enthusiasm and confidence finally persuaded him to agree. The enormous undertaking was planned in a single week and went ahead under the code name 'Market Garden'. The British 1st Airborne Division was to lead the attack, supported by Allied ground forces with American airborne troops responsible for taking the bridges south of Arnhem.

The first strategic error was probably in the choice of drop and landing zones at Arnhem. Four main ones were chosen (there were two others, and a supply dropping point) some 6 miles outside the city. The RAF had advised Montgomery that the Division should not be landed close to the town because of the danger from anti-aircraft fire. But these zones proved to be too far from the Arnhem road bridge, which was the main objective. In the event, only some 700 men out of the whole 10,000 reached the bridge. In balancing the relative risks, Montgomery should have overruled the RAF. As it turned out, the anti-aircraft fire was surprisingly light during the first two major landings.

Montgomery's second error was to agree with the RAF's plan to spread the drop over three days, instead of making two trips in one day. The RAF had not enough aircraft to do the whole operation in one drop and wanted to minimise the strain on aircrew and ground staff that would result from making the same flight twice in one day. The element of surprise was of course lost after the first landing; and the weather then deteriorated. So the second landing was delayed by several hours, and the third by three days. Twenty-four hours after the first landing the Germans had successfully organised their defences.

It was also a mistake to overlook the danger prevented by the fact that the only road available for Second Army's advancing tanks was clearly visible to the German artillery. The Dutch officers who were attached to the General Staff, including Prince Bernhard, could have warned them about this, but they were excluded from the planning sessions. Had they been there, they would also have reported that two SS Panzer Divisions were refitting near Arnhem. Perhaps the Dutch were excluded because of fear of possible leakage of information to the Dutch Nazis.

The need for good radio communication was underestimated; during the whole Battle of Arnhem the equipment proved totally inadequate. It was not suitable for the terrain and was not powerful enough, either for ground-to-ground or ground-to-air communication.

Another factor contributing to the Allied defeat could perhaps not have been foreseen; in spite of their exhausting retreat across Europe,

Summing Up

the German Army responded to the Allied landings with astonishing resilience, imagination and speed. Montgomery gravely underestimated the quality of his enemy.

But perhaps the most serious error was to land the British invasion force right into the middle of the greatest concentration of German forces on the Western Front.

Montgomery persisted in his choice of landing at Arnhem in spite of a succession of detailed reports from the active and reliable Dutch Underground, and from Allied air reconnaissance. To Model and his staff it had seemed impossible that the British should not have known what was going on behind the German lines. To Montgomery it was unthinkable that his ambition to be first in Berlin should not be fulfilled. Against the prospect of a dazzling victory, and an end to five years of war, he persuaded himself that the risk of enormous casualties was worth taking. His powers of self-deception did not fail him, even in defeat. Thousands of men were killed or wounded, and thousands more taken prisoner; hundreds of aircraft were lost; and the Allied armies were unable to advance into Germany for another four months.

Yet Montgomery's comment was that the operation was '90 per cent' successful; he remained, he wrote, 'an unrepentant advocate of "Market Garden"'.

7

Life After Arnhem

The morning after I got back to England, Wing Commander Lillywhite, commanding officer of the airfield, called me into his office. He told me that I had been awarded the Military Medal for 'bravery in the field', and that I was to be decorated by the King at Buckingham Palace. He then mentioned a slight problem – my unmilitary bearing and sloppy uniform. 'It would not fit the occasion. In fact,' he said, 'it's a tragedy that the only man on the airfield to be decorated is a scarecrow like you.' He suggested that after my two weeks' survivor's leave I should report to the Guards Brigade at Wellington Barracks, near Buckingham Palace, to be smartened up and to learn the correct drill for the investiture.

I had never heard of the Military Medal and could not imagine how anyone in London could have known about what I had done at Arnhem a few days earlier. Later I learnt that it was Captain Ogilvie – the 'Captain Z' of *Arnhem Lift* – who had sent a radio message from the HQ at Oosterbeek recommending me for the award. I was deeply upset to learn that he had been drowned crossing the Rhine: he was a really good man, warm-hearted, and with a great sense of humour. I left Commander Lillywhite in a happy daze – ahead of me I had my survivor's leave and several more weeks in London as well.

And now, wherever I went, friends asked me what had happened at Arnhem. I got so bored repeating the story that I thought I had better write it all down and get it duplicated. The girlfriend I was staying with became most enthusiastic about the idea, but I soon got tired of it. She was determined, however, and devised a regime by which I was only allowed the necessities of my life – love and food – when I had completed a few typed pages. I tried to fool her by enlarging the margins and increasing the spacing of the lines, but she always found out. In this way we completed a 150-page report in two weeks. She was still enthusiastic and thought it should be published.

For this I had to submit my story to my commanding officer, Colonel Murray. He was outraged. 'You should be ashamed of ourself!' he said, 'No Britisher would ever have let his comrades down by writing stuff like this. It lets down the whole regiment! I won't even pass it to the War Office. I forbid you to contact either a publisher or anyone from the press.' He concluded, 'Sergeant Haig, I am very disappointed in you – you of all people, a holder of the Military Medal.'

But my determined girlfriend sent her copy to the War Office without telling me. It was passed for publication, with the proviso that anything in it that might indicate that the author was German had to be cut out. But I had no idea how things were going to work out because Colonel Murray – obviously too ashamed to keep me under his command – had me transferred to a Glider Pilot Regiment unit in India.

Before I left (at the very end of 1944) I had to attend the investiture at Buckingham Palace. There a tall, elderly, most elegantly turned-out major-domo in a frock coat inspected us to make sure we were properly dressed. I was escorted through an antechamber to a large arched double door that led into the ballroom where the investiture was taking place. I was gently pushed through the door.

As smartly as I could, I marched the long distance across the parquet floor, my heavy army boots echoing through the hall. I halted, stamping my feet noisily, and saluted in front of the King, just as the Guards corporal at Wellington Barracks had taught me. With the King were the Queen and their two young daughters, the Princesses Elizabeth and Margaret. They all shook my hand. Then an equerry handed the King a large silver medal on a velvet cushion. He pinned it to my chest and congratulated me. 'And where do you come from?' he asked me, and, before I could think it over, I answered, 'Potsdam, Sir!' The King stammered out something I couldn't understand. Then, after an awkward pause while his jaw moved without a sound coming out of his mouth, he continued, 'Arnhem – a terrible tragedy – so many men. But I am sure you are keen to get out there again soon.' I was mesmerised by the struggle the poor man had to get his words out, and before I knew what I was saying, answered, 'No, Sir. I only just got safely back!'

From the look in his eyes I realised the interview was over. I stamped my feet, saluted and marched out of the hall. Later my friends told me that the Queen had stared fixedly at me, plainly wondering how a German had come to be in her drawing room.

I was very upset about being sent to India: I wanted to be in Europe when the war was won and Nazi Germany defeated. But once I was there, I found it fascinating and, to while away the boredom – we had no aircraft to train with – I started to write. (Later some of what I had written then was published under the title *Indian Route March*).

After a few months of idleness, in the excruciating heat of the spring, several of us were sent to the famous officers' training college in Poona. I completed my course successfully, except for one highly embarrassing incident which might easily have resulted in my dishonourable discharge from the college.

At six o'clock every morning we rehearsed the most complicated manoeuvres on the parade ground in preparation for the march-past in front of a senior officer who 'took the salute' each week. I always found a place in the middle of the ranks so that I could just follow the others in their various movements while thinking about something more interesting. One fine Sunday morning the regimental sergeant-major told me I was to have the honour of leading the parade. And it was to be a special honour, because a visiting American general who had been in 'Market Garden' was taking the salute and said he wanted to meet me. Before he had finished I blurted out, 'Oh no, Sir! I couldn't possibly. Please choose someone else.' The RSM simply ignored my pleading with a brisk 'There's no need to be modest, my good man. This is an order.'

The bugle sounded the fall-in, and the RSM shouted, 'Marker ... Out!' I marched out as smartly as I could and the rest followed. After that my mind went blank. Apparently I led the parade into a left wheel and then three right wheels, so that it cut off its own tail and was soon in complete disarray. The rest of the cadets were doubled up with hysterical laughter and couldn't even hold their rifles upright; they were mercifully released by the bugler desperately sounding the 'Retreat'.

I was summoned by the commanding officer. He addressed me in a quiet, resigned voice: 'What you did is so disgraceful that it has no precedent in the 200 years' history of this college. We have not yet been able to decide on a suitable punishment, but you may be sure that you will hear from me in due course. In the meantime, I never want to see you anywhere near the parade-ground again.'

Apparently no suitable punishment was ever found, for I never heard any more about it. But from that time on I was excused every parade.

I passed my course, but at the end I did not wait for my official commission. Instead I went off to Calcutta. There, over a few beers in the Grand Hotel, I got into conversation with a naval officer who noticed the wings on my uniform and asked me about Arnhem.

He was Tony Clarkson, one of the editors of *Phoenix Magazine*, published for the English-speaking troops of South-East Asia Command. He showed me the current issue, which prominently featured extracts from a book published in England in January, *Arnhem Lift: Diary of a Glider Pilot*, by an anonymous writer. I quickly read some of the passages. It was my book! The title and publication date had been decided after I had left for India – this was the first I knew of it. Tony Clarkson was delighted; he had my photograph taken and published it in the next issue with a highly dramatised story about how they had found the anonymous author of *Arnhem Lift*.

A week later Tony Clarkson handed me a little red pass certifying that 'Lewis Haig', correspondent for *Phoenix*, should be given every assistance in obtaining information, transport, billets and finance. It was signed 'Louis Mountbatten, Commander-in-Chief, South-East Asia Command'.

Life After Arnhem

After I had done features all over India on religious festivals, primitive jungle tribes, prostitution and British memsahibs, Tony told me that he was going to send me over to Burma, Malaya, Siam (now Thailand), Indonesia and Singapore. I was to write features and take photographs. When I said that I had only ever taken a few snaps with a Box Brownie years ago in Germany, he said it didn't matter. 'It will be much easier with the Leica you'll be taking with you. I'll be satisfied with one or two printable pictures in each reel of 36.'

So off I went, first to Rangoon, where I wrote a story about Major-General Orde Wingate,[11] the leader of the Chindits, a British irregular force that had operated behind Japanese lines in the mountainous Burmese jungle. Then I did a feature on Penang, the forces' dream holiday island, just off the Malayan mainland. From there I flew to Singapore and on to Bangkok on the heels of the retreating Japanese.

On my last assignment for *Phoenix* I was sent to Indochina (now Vietnam). The French had occupied a large part of the country around the capital, Saigon, but in the north, Hanoi was held by a people's army, under their Moscow-trained leader, Ho Chi Minh. They had fought the Japanese and were now fighting for their own independence from France. At this time there was an uneasy truce between Saigon, in the south, occupied by the Allies, and Hanoi in the north, ruled by Ho Chi Minh. I made friends with one of his senior advisers, a German communist, who arranged for me to interview Ho – the first Western journalist to do so. Ho Chi Minh was surprisingly small and delicate; he answered every one of my questions courteously and to the point. To my astonishment he advocated cooperation with the Western Allies, saying, 'After all, we are both fighting the Japanese; though my people will never revert to their position under French colonial rule.'

I wrote a long feature advocating cooperation between the two regimes, but it was censored by the French authorities. France laid claim to the whole of Indochina, a short-sighted policy that eventually resulted in the long and bloody Vietnam War.

---

11 Wingate had been killed in an air crash in 1944.

After demobilisation I got a job on the *Sunday Express* in Berlin; later, Tony Clarkson, by then an editor with Odham's Press, sent me all over Germany to write for the weekly illustrated *John Bull*, and for *Country Life*. I used some of these articles for a book about the rise of Nazism called *Follow my Leader* (republished in paperback by Spellmount with the title *Ein Volk, Ein Reich: Nine Lives Under the Nazis*). This book tells the story of the Nazi regime – how it happened and why it happened – through interviews with ordinary Germans, instead of with their leaders. Later I wrote, edited and translated several other books, and in 1950 I started Primrose Film Productions with an office off Primrose Hill, near Regent's Park in London.

I now have two homes, one in England and one in Norway – my wife is a Norwegian painter – and we travel a lot, especially to Germany to visit what is left of my family. But England is my home, and if someone asks me what I am – German, Norwegian, Jewish or British – I answer, 'I'm an Englishman.'

Index

Abbey Arts Centre, London 138
Ack-Ack 194, 209, 235
Adenauer, Konrad 17
Anstiss, Herbert 50
Apeldoorn 275
Arbeitsdienst 264
Ardennes 268
Army Air Corps 69, 71, 93, 94, 175
Army units, Allied 123, 176
   1 Airborne Corps 181
   1 Airlanding Brigade 181
   1 Parachute Brigade 181, 279
   1st British Airborne Division 167, 168, 180, 181, 289
   4 Parachute Brigade 181
   21 Army Group 168, 180
   82nd US Airborne Division 167
   101st US Airborne Division 167
   165 Company 61
   Army Air Corps Glider Pilot Regiment 175
   British Second Army 180
   King's Own Scottish Borderers 202
   Polish 1 Independent Parachute Brigade 181
   XXX Corps 180–1
Army units, German 123, 176
   II Parachute Corps 280
   9th SS Panzer Division 275, 283
   10th SS Panzer Division 275, 279, 281
   191st Artillery Regiment 286
   Afrika Korps 265–6
   Army Group 'B' 268, 272
   Fifteenth Army 280
   Panzer Reconnaissance Unit 265
   Sixth Army 266
   SS Panzer Grenadiers 209
Arnhem Lift 9, 98, 115–7, 126, 136, 139, 162, 183, 293, 296,
Arnhem, Netherlands 75, 271
Australia 57, 174

Bangkok, Thailand 121–3, 297
Barn Theatre, Guildford 52–3, 94, 128
Behr, Major Winrich H.H. 139, 163, 263, 269, 271–3, 277
Belgium 244, 265, 268, 271
Berkeley Square, London 48
Berlin, Germany 10, 20, 25
Bernhard, Prince 140, 187–8, 290
Bevan, Nye, MP 12, 38
Bideford, England 59, 174
Bittrich, Obergruppenführer Wilhelm 273–4, 278, 282
Block, Otto 27
Block, Peter 150

BMW 19, 32, 35, 43, 47, 169–70, 172, 274
Bormann, Martin 267
Brandon, Bill 43–4, 49
Bray, Reginald 52
Bren gun 82, 199, 203–5, 242, 244
Brinkmann, Sturmbannführer Heinrich 275, 282
Browning, Lieutenant-General Frederick A.M. 168, 181, 257
Brownshirts (SA) 31–2, 170
Brussels 91, 258, 269
Buckingham Palace, London 93, 96, 293–4
Burcell, John 52

Calcutta, India 115–9, 125, 296
Canada 56–7, 174
Cirencester, England 61–2
Clarkson, Tony 116, 120, 296, 298
Cologne, Germany 17
Coster, Ian 115–6
Cowley, England 39, 172
Cudlip, Hugh 116
Curling, Lieutenant 72–4
Czechoslovakia 264

Dakota 194, 235
de Havilland Tiger Moth 176
Deelen 181
Dempsey, Lieutenant-General Miles 180
Denham 72–3, 176
Doetinchem 273, 274, 276
Deutsche Bank 19
Devon, England 58, 61, 174
Donnington, England 65–8
DUKWs 258
Dunkirk, France 61, 174, 252, 265, 283
Dutch National Socialist Party 224
Dutch underground 231, 277, 291

Ede 276
Edgar, Curt 127
Ein Volk, Ein Reich 10, 298

Eindhoven 277
Eisenach, Germany 32
Eisenhower, General Dwight 179, 180, 289
Elizabeth, Princess 97, 294
Enemy Alien 55–7, 59, 61, 115, 173, 175
Engelman, Judge 34

First World War 17, 19, 25, 62, 71, 75, 166, 172, 263
FitzGibbon, Constantine 58
Fitzroy Square, London 11, 39
Focke-Wulf 189, 209, 235
Follow My Leader 10, 133, 298
Frankfurt 128, 269
Frost, Lieutenant-Colonel John 279

Galtür, Austria 135
Gatsby, Peter 55
Geneva Convention 83, 205, 265
George VI, King 96
German National Party 264
Gertud, Victoria 15
Göring, Herman 17
Gough, Major Freddie 52, 282
Gough, Michael 52, 282
Graham, Sergeant 85, 207, 212, 214, 223, 238, 241
Grave 167, 180, 186
Griffith, Hubert 11
Groote Heide 278
Gulmarg, India 110
Gussow, Karl 18

Hackett, Brigadier John 212, 229
Hagen, Anne Mie 137, 143, 159, 162
Hagen, Carl (Opapa) 18, 19, 138, 147
Hagen, Emma 18
Hagen, Hans Peter 26
Hagen, Karl-Victor (KV) 26, 35, 127, 128, 134, 135
Hagen, Karoline (Carla) 10
Hagen, Dr Louis 10, 150
Hagen, Louis (Büdi) 162, 163, 171, 271, 278, 281, 286

Index

Hagen, Louis George 47, 48
Hagen, Nina Katheriena 26, 27, 35, 134, 170
Haig, Lewis (aka Louis Hagen) 75, 116, 177, 296
Halifax 194
Hallstein. Professor Doctor Walther 269
Hanoi, Vietnam 123, 297
Harder, Sturmbannführer 282
Harmel, Brigadeführer Heinz 279
Hartenstein hotel 263, 271, 281, 282, 287
Hartenstein pocket 286
Harzer, Obersturmbannführer Walter 274

Heelsum 181
Hexenkessel 285
Highgate, London 169
Hitler, Adolf 264
Ho Chi Minh 123, 297
Home Office, England 56, 59
Horrocks, Lieutenant-General Brian 180

India 99, 101, 104, 106, 110, 113, 116–21, 126, 294–97
Indian Route March 133, 295
Iron Cross 31, 265

Japan 136
Jodl, Colonel-General Ferdinand A.F. 267
Johannahoeve 280
Jungfernsee, Germany 27, 149, 169

Karachi, Pakistan 101
Kashmir, India 110
Keitel, Field Marshal Wilhelm 267
Kiel, Germany 25
Kiepenheuer, Woelfchen 128
Knight's Cross 265
Koch, Carl 138
Kotter, Wilhelm 19

Kraft, Sturmbannführer Hans 273–5, 278, 284
Krebs, General Hans 263, 268, 271–3, 277
Kremer, Ans 139, 140
Kurfürstendamm, Berlin 20, 127

Lee, Jennie, MP 38
Levy, Abraham (A.E.) 17, 18
Levy, Albert 17, 21
Levy, Herman 17
Lillywhite, Wing-Commander 93, 293
Lippert, Standartenführer Michel 279
Loewy, Victoria (Vicky) 21
London, England 10, 11, 35, 38, 45, 47–9, 51–3, 67, 72, 93–6, 101, 133, 151, 162, 169, 172, 293
Lore, Woelfchen 128, 129
Louvain 91, 258

Maas river 167, 180
Manet, Édouard 20, 152
Margaret, Princess 97, 294
Matz 25, 26
McFadyean, Sir Andrew 35, 56, 67, 172–3
Medawar, Jean 96, 121, 177
Medawar, Peter 163, 177
Messerschmitt 189
Meuse river 180
Michaelis, Anthony 57
Milch, Field Marshal Erhard 267
Military Medal (MM) 93–4, 293–4
Milroy, Dido 94, 126
Milroy, Vivian 94, 138, 163
Model, Field Marshal Walter 268
Mohrenwitz, Dr Lothar (Mumpitz) 11
Mollers, Sturmbannführer 286
Monet, Claude 158
Monnet, Jean 269
Montgomery, Field Marshal Sir Bernard 140, 180, 289
Mountbatten, Commander-in-Chief Lord Louis 116, 296
Mueller, Hank 47

Mulberry harbour 179
Murray, Colonel 94, 98, 212, 229, 294

NAAFI 66, 259, 271
Nagin Lake, India 110
Nazi(s) 10, 31, 35, 37, 41, 56, 78, 83, 120, 128, 131, 135, 138, 171, 173, 197, 225, 230–1, 264, 290, 298
New York 10, 21, 35, 134–7, 151
Nice, France 19
Nijmegen, Netherlands 91, 167, 168, 180, 231, 242, 256–8, 274, 275, 277, 279–82
Norway 137, 162, 298
Notting Hill, London 58, 174

Oeynhausen-Sierstorff, G.N. 263
Ogilvie, Captain ('Captain Z') 82, 84–6, 88–91, 94, 202, 293
Oosterbeek, Netherlands 81, 94, 139, 140, 199, 263, 271–6, 279, 281, 282, 284–7, 293
Oppenheim, Salomon 17, 18
Oppenheimer, Joseph 29
Oslo, Norway 162
Ostier, Robert 123
Owen, Colonel Frank, MP 116
Oxford 35, 39, 43–4, 48–9, 51, 55, 64, 68, 72, 75, 96, 117, 172, 175, 177

Pannerden 279
Paris 35
Penang, Malaysia 120, 297
Philippi, Katherina 18
Phoenix Magazine 115–6, 119, 120, 123, 125, 296, 297
Pilot Press 99
Pioneer Corps 59, 61, 174
Plant, Jimmy 200–1
Plummer, Dick 126
Poland, Polish 179, 264–5
Poona (Pune), India 113, 115, 295
Potsdam, Germany 31
Pressed Steel Company 35, 39, 47, 55, 172

Prestcold Refrigerators 43, 48, 173
Primrose Hill Film Productions 95, 138, 162, 298
Primrose Hill, London 298

RAF Denham 72–3
Rangoon (Yangon), Myanmar 120, 297
Red Cross 77, 83, 186, 205–6, 243, 284
Reiniger, Lotte 138–9
Renoir, Pierre-Auguste 20, 138, 152
Rhine River 75, 77, 88–91, 94, 167, 177, 180, 184, 230, 244, 247, 251–5, 281, 285–7, 293
Rhineland, Germany 17
Royal Army Medical Corps (RAMC) 68, 126
Royal Army Service Corps (RASC) 228
Royal Electrical and Mechanical Engineers (REME) 64, 175
Ruhr 181, 277

SA 31, 34–5, 170, 171, 190
Saigon (Ho Chi Minh City), Vietnam 123, 297
San Francisco 136
Santa Margarita, Italy 135
Scheldt 280
Schloss Lichtenberg Concentration Camp 31–2
Schmundt, Major-General Rudolf 267
Seine river 167, 268
Self-propelled gun (SP) 79
Sherman tank 266
Singapore 120, 121, 297
Six-pounder 280
Skalka, Hauptsturmführer Doktor 284
Spandau machine gun 83, 204, 236–7, 250
Spindler, Sturmbannführer Ludwig 275, 283, 286
Spitfire 64, 78, 107, 189
St Elizabeth Hospital 275, 284
Stalin, Josef 27, 264

Sten gun 80–1, 85, 90, 196, 199, 212–3, 232, 248, 250, 253–4
Stirling 194
Strutt, Lieutenant 83, 87
Sunday Express 126, 130, 133, 298

T34 tank 272
Taylor, Jean 39
Terborg 276–7
The Adventures of Prince Achmed 138
The Mark of the Swastika 133
Theresienstadt Concentration Camp 130
Threshold Theatre, London 58
Tiger Moth 72, 176
Tiger tank 276, 283–5,
Tobruk 265
Torgau 31, 170, 190
Travemünde, Germany 25–6
Typhoon 78, 188–9

United States 136–7
Urquhart, Major-General Roy 181
Ustinov, Peter 52
Utrecht 276

Villa Carlshagen 21, 22, 149, 152
Villa Hagen 29, 30
von Allworden, Sturmbannführer 282, 286
von Kluge, Field Marshal 268
von Manstein, Field Marshal 267

von Metzsch, Oberleutnant 272
von Osterode, Baroness Mausi 129
von Paulus, Field Marshal 266
von Rundstedt, Field Marshal Gerd 268, 272, 286
von Tettau, Generalleutnant Hans 278–9, 284
von Wassermann, Georg 59

WAAF 259
Waal river 167, 180, 278
Wade, Vic 200
War Office, England 47–8, 58, 60, 63, 68, 98–9, 175, 294
Warburg, Sigmund 172
Warrack, Lieutenant-Colonel Graeme 233
Wellington Barracks, London 93–6, 293–4
Wellington Square, London 39
Western Front (First World War) 26, 265, 291
Wettsetin, Julia 18–9
Windrush airfield 271
Wingate, General Orde 120, 297
Wolfheze, Netherlands 75, 140, 180–1, 184, 188, 235, 241, 265, 271
Wolfsschanze 266–7
Wolters, Lieutenant-Commander Arnoldus 231

Zehlendorf, Berlin 127

You may also enjoy …

978 1 80399 024 8

The Battle of Arnhem has acquired a near-legendary status in British military history as an audacious plan to land paratroopers into the Netherlands and spearhead an attack against the German-held Ruhr. Using first-hand accounts, maps and detailed timelines, historian Chris Brown explores the unfolding action of the battle and puts the reader on the front line. If you truly want to understand what happened and why – read on.

You may also enjoy …

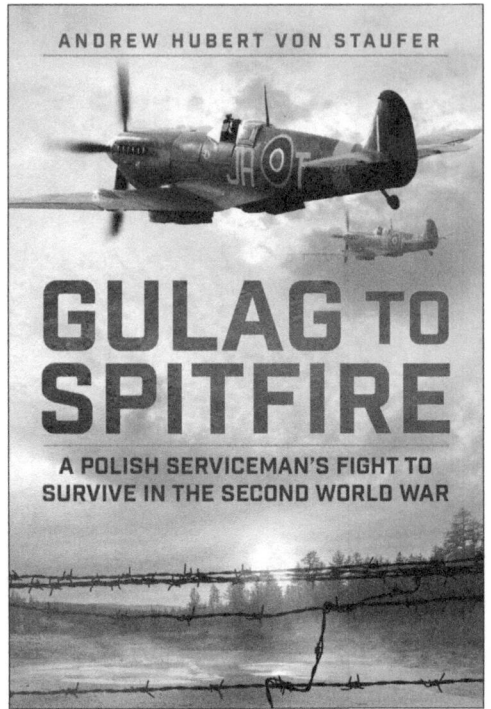

978 1 80399 521 2

Stalin is quoted as saying 'One man's death is a tragedy, a thousand deaths is a statistic'. *Gulag to Spitfire* is the story of a man who was determined to be neither. In this moving tale of endurance against all odds, Andrew Hubert von Staufer traces his father's footsteps from the gulags of Siberia to flying Spitfires in air battles against the Luftwaffe.

The destination for history
www.thehistorypress.co.uk